# Discovering the Desert

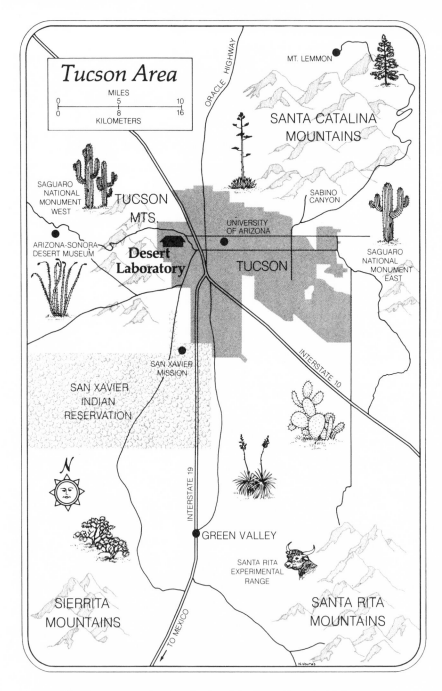

The Tucson area, showing the site of the Desert Botanical Laboratory.

# Discovering the Desert

## Legacy of the Carnegie Desert Botanical Laboratory

WILLIAM G. McGINNIES

THE UNIVERSITY OF ARIZONA PRESS
TUCSON, ARIZONA

*About the Author . . .*

WILLIAM G. MCGINNIES, an internationally recognized authority on arid lands, first became acquainted with the desert in 1918, when he moved from Colorado to Tucson to attend the University of Arizona. It was later, as a member of the faculty of the College of Agriculture, that he developed a close association with personnel at Carnegie's Desert Botanical Laboratory. Soon after the Laboratory had closed in 1940, McGinnies left Arizona, but he returned to the University in 1960 to become head of tree ring studies and arid land activities. He was founder and director of the Office of Arid Lands Studies, which he continued to serve as a consultant after his retirement.

Third printing 1987

THE UNIVERSITY OF ARIZONA PRESS

This book was set in 11/12 Videocomp Palatino.

**Library of Congress Cataloging in Publication Data**

McGinnies, William Grovenor, 1899-
    Discovering the desert.

    Bibliography: p.
    Includes index.
    1. Desert flora.   2. Desert ecology.   3. Desert
ecology—North America.   4. Desert ecology—North America.
5. Carnegie Institution of Washington. Desert Botanical
Laboratory—History.   I. Title.
QK938.D4M37        581.5′2652         81-1554

ISBN 0-8165-0719-8                     AACR2
ISBN 0-8165-0728-7 (pbk.)

## To Forrest Shreve

To know the desert involves an acquaintance with all its aspects, and all its physical features, as well as all of the animals and plants that have learned how to find in it a congenial place to live. The most significant lesson that the desert dweller can learn from a familiarity with its plant and animal life is to regard himself not as an exile from some better place, but as a man at home in an environment to which his life can be adjusted without physical or intellectual loss.

Forrest Shreve, *The Cactus and Its Home*

# Contents

## Tables

# Illustrations

## *Figures*

## Maps

# Foreword

*The greatest service which can be rendered any country
is to add a useful plant to its culture. . . .*

THOMAS JEFFERSON

In his Lindbergh lecture, "Science and Technology in a Conserving Society," Frank Press said that of the approximately 350,000 plant species described by botanists, only about 3,000 have been tried as sources of food and other useful materials. We cultivate only about 100 plant species on a large scale, about 90 percent of our food coming from only a dozen or so crops. Thus the urgency of identifying plants that can survive in, or can be adapted to, areas where it is difficult to cultivate today's conventional crops—the arid regions, for example, which constitute almost 19 million square miles, over one-third of the earth's land surface.

Emanuel Epstein wrote in *Genetic Engineering of Osmoregulation* that terrestrial plants in arid and semi-arid regions face severe problems of water economy, because the salinity of the soil gives rise to osmotic withdrawal of water, and hot, dry atmospheric conditions cause excessive loss of water due to transpiration. He went on, "Yet despite these precarious conditions, the arid and semi-arid regions are among the most promising ones to turn to in our quest to increase the production of food, fiber, chemicals, and biomass for energy. The relatively unleached soils of these regions are often inherently fertile, the growing season is long, temperature and light intensity are high, and the atmospheric humidity is low, reducing disease problems. All these features favor agricultural productivity if water and salinity problems can be solved. . . ."

Today, we stand at the threshold of solutions to some of the most difficult problems that confront mankind—the enhancement of the productivity of plants in stress environments —as students of the genetics and molecular biology and of the physiological ecology of plants join forces. Once we understand how certain plants adapt to drastic changes and water availability, the techniques of genetic engineering can be brought to bear, assuring new generations of temperature- or salt-hardy plants and enabling us to utilize marginal lands for energy crops.

The happy marriage of ecology with biochemistry and physiology may be illustrated with one example from today's Carnegie Department of Plant Biology. Björkman, Berry, and their colleagues have made substantial progress in understanding how certain plants can adapt to drastic changes in temperature and water availability. One such plant, commonly found as a highway planting in California, is the shrub called oleander. It is found throughout the state, from the desert regions to the mountains, and seems to thrive on neglect. What is happening when identical clones of oleander thrive both at temperatures approaching 120°F and at temperatures near 60°F?

It turns out that there are two major differences between the plants grown at hot and cool regimes. First, those plants from the cool regime can photosynthesize much more effectively at low temperatures than can those from the warm regime. Indeed, the plant from the cool regime achieves more than double the rate of photosynthesis of the warm regime plant when both are tested near 65°F. It simply has a greater amount of one particular limiting catalyst—almost two and one-half times as much. Otherwise, there are no significant differences in the biochemical machinery for making additional biomass.

The other major difference between plants grown at warm and cool regimes is found when the plants are investigated at high temperature. Here the situation is reversed, the photosynthetic rate of the warm regime plants far surpassing that of the cool. Again, the difference turns out to be relatively simple: the membranes from the high-temperature-grown plant are far more stable than those of the low-temperature-grown shrubs.

It is only a small oversimplification to say that they develop membranes with a higher melting temperature. Even more surprising is the finding that mature plants can adjust to dramatic temperature changes. Thus, if a mature high-temperature-grown plant is moved to the low temperature regime, the melting temperature of the membranes decreases (they must remain fluid at lower temperatures) and the amount of the limiting enzyme increases. These changes are rapid, and, within a week or so, the plant is completely adapted to its new regime. With the reciprocal change—a low-temperature plant moved to high temperature—the reciprocal changes in membrane properties and enzyme level occur.

It is clear that not all plants are as adaptable as oleander, however. The plant breeder now has at hand simple tools for screening for such adaptability. The membrane properties can be studied in the field; in seeking such desirable traits in crop plants, they need no longer be grown through whole generations. They can be screened quickly at a young age, and the geneticist and the breeder know exactly what to look for.

In his paean to the symbiosis of humankind and the earth, the *Wooing of Earth,* Rene Dubos wrote "We cannot escape from the past, but neither can we avoid inventing the future."

Björkman and Berry, and their colleagues in laboratories around the world, are "inventing the future"—and it promises to be a bright future indeed. But all of them are keenly aware that today's contributions rest squarely on the shoulders of the giants of the past. For nearly four decades, the geographic frontier of research on desert plants was Tucson, and the intellectual frontier was found in the minds of Forrest Shreve, Frederic E. Clements, Herman Spoehr, and their colleagues in the Carnegie Desert Laboratory.

Through William McGinnies' book *Discovering the Desert: Legacy of the Carnegie Desert Botanical Laboratory,* we can now relive the story of what was, for a significant period, an intellectual oasis in the desert.

JAMES D. EBERT
*President*
*Carnegie Institution of Washington*

# Preface

THE CARNEGIE DESERT BOTANICAL LABORATORY was the first research institution in the world to be devoted entirely to desert problems. It was established by the Carnegie Institution of Washington to "study the life history of plants under desert conditions with special reference to absorption, storage and transpiration of water, to enable agricultural experiment stations to make practical application of this knowledge." I believe this objective was accomplished. Scientific recognition has been given to much of the research, but a great deal of basic information has not been fully utilized. Many findings of the Desert Laboratory can still be of practical use to agricultural experiment stations, to other professional groups, and to nonspecialists.

The far-reaching influence of the Desert Laboratory is one measure of its success. Its investigations stimulated subsequent desert research activity throughout the world. The organization of an Advisory Committee for Arid Zone Research under the auspices of UNESCO in 1951 was an outgrowth of this increasing interest. That committee has been instrumental in the development of many national arid land research institutions, which in turn have started global programs for restoring depleted lands and improving the use of arid land. The world owes a debt of thanks to the pioneering scientists at the Desert Botanical Laboratory. Their work has been the solid foundation on which others have built to expand understanding and appreciation of the desert and its environment.

My twofold purpose in writing this book was to credit the high level of scientific accomplishment by Laboratory personnel and to bring together in readable form the results of their work for the increasing number of people who are interested in deserts and their unique vegetation.

The vast store of information on the Desert Laboratory findings necessitated a process of selection and interpretation. In selecting, I felt it was appropriate to express my deep appreciation of the men and women who devoted so much of themselves for varying periods of time from the establishment of the Laboratory in 1902 until it was turned over to the U.S. Forest Service in 1940. The Laboratory's history, as recounted here, therefore includes personal reflections and a few anecdotes that may vivify the reader's understanding of the conditions under which early desert studies were performed.

Although the Laboratory lacked many latter-day improvements of equipment and technique, its studies of the Sonoran Desert were the most comprehensive of any desert studies performed until the time of its closing. Despite difficult working conditions, climatic discomforts, and almost impassable roads in Arizona and Sonora, Laboratory personnel made discoveries of lasting value. The modern traveler can still rely on these early studies in botany, physiography, soils, and climate to illuminate what he or she sees while speeding over the paved highways of the region.

The Laboratory studies included climate in general—rainfall distribution, temperature, and such matters as temperature inversions and soil moisture relationships—and investigation of the plants themselves and how they are able to live in a country with little rainfall.

Significantly, much of the work at the Laboratory pertained to nonagricultural lands, in contrast to the work of collegiate federal experiment stations. Desert Laboratory studies were likely to be in peripheral areas where it was more difficult to obtain information on the relationship of plants to soil conditions. Early soil surveys generally stopped at the edge of the terrain beyond which surveyors believed that ordinary agricultural practices must cease. Much of the work of agricultural science on soil moisture relations had been done with fairly homogeneous soils, without the problems caused by shifting sand, rocky depths, or calcareous layers, with their wide range of effect on soil conditions. For these reasons, the Desert Laboratory studies are of value even though, in some instances, their answers are not complete.

The climatological contributions of Desert Laboratory scientists are of special interest because they differed from the usual Weather Bureau station studies located in cities, often not the best places for observing climatic conditions. These investigations extended into areas between the cities, with rain gauges in locations far removed from urban areas. Specifically, they established a set of moisture stations between Tucson and the Gulf of California, and between Tucson and Yuma, located to represent a gradient that could be interpreted in terms of the moisture conditions of a transect. Although these precipitation stations were operated for only a short period, they gave a great deal of information useful in the interpretation of long-time records obtained from regular stations.

Desert Laboratory workers also observed the characteristics of rainfall: the seasons in which rain occurs, the amounts and distribution, and whether winter or summer rainfall is more regular from year to year and from place to place.

Information about the relationship between rainfall and soil moisture, and between runoff and percolation has become recognized as not only valuable but interesting to home gardeners, to arid lands residents, or to any residents of the planet who are curious about the multifaceted relationships of desert plants and the growth and survival of desert vegetation.

After considerable thought, my decision was to give measurements in English units with metric equivalents, for the benefit of the greatest number of readers. Because exact equivalents often resulted in awkward values, some liberties have been taken in the expression of equivalents. For example, where a general reference stated soil moisture penetrated 3 to 6 feet, this was assumed to be approximate, and the metric equivalent is given as 1 to 2 meters. Similarly, when elevations are expressed in round numbers (i.e., 3,000 feet), the approximate equivalent is used—3,000 feet becoming 900 meters. In areal extent the same liberties were taken. For general precipitation figures the conversion was based on 25 millimeters or 2.5 centimeters equal to 1 inch; thus, 200 millimeters (20 cm) equals 8 inches.

Plant names posed problems until it was decided to use common names in the text and to provide common name-scientific name lists in the appendix. For scientific names, Shreve and Wiggins (1964) were the authority followed, despite some latter-day changes in nomenclature. Common names as often as possible were based on common usage in the Sonoran Desert area. Reviewing authorities were in accord on most of these names, but, where no agreement was reached, I selected what seemed to me most appropriate.

WILLIAM G. MCGINNIES

# The Establishment
# of the Desert Laboratory

## Beginning of an Era

ON JUNE 28, 1902, the newly appointed advisory committee for botany of the Carnegie Institution of Washington, D.C., recommended "the establishment and maintenance of a desert botanical laboratory for the purpose of ascertaining how plants perform their functions under the extraordinary conditions existing in deserts."

The Committee determined that there should be established at some point in the desert region of the southwestern United States a laboratory for the study of the life history of plants under desert conditions, with special reference to the absorption, storage, and transpiration of water. Although there were many botanical laboratories in the humid portions of the temperate regions, as well as several marine laboratories and tropical laboratories devoted in whole or in part to botanical research, a desert botanical laboratory existed nowhere in the world.

Yet, the phenomena presented in the adaptations of plants to desert conditions were felt to be the most interesting and significant—from an evolutionary point of view—of any in the whole realm of botany. The economic basis for the establishment of such a laboratory was the enormous development of population and industries that was bound to take place in this arid region during the next hundred years. The basis of that development would be agriculture, both with and without irrigation.

In the early twentieth century, little was known about the peculiar fundamental processes of plant growth under the unusual conditions surrounding plant life in desert regions. The investigations proposed were of so general a character, so expensive, and so difficult that no agricultural experiment station had yet undertaken them, and there was no prospect that any station would do so.

Thus, on December 3, 1902, the Board of Directors of the Carnegie Institution formally appropriated $8,000 for the purpose of establishing a desert laboratory and maintaining it for a year. By the end of December, W. A. Cannon, of the New York Botanical Garden, had been selected to become Resident Investigator in charge of the laboratory, at a salary of $150 per month.

### The Site and the Building

Although the members of the Advisory Board had visited most of the more marked desert of the country within the preceding twelve years, it was necessary for them to make a more systematic tour of these deserts in order to gain greater comparative knowledge of the aspects of the vegetation and to select a locality with the greatest advantages and facilities for the proposed work.

The principal requirements established for a laboratory were a distinctly desert climate and flora; a flora rich and varied as possible, while still of distinctly desert character; ready accessibility; and habitability.

Tucson met these qualifications and was chosen as the site. The climate was of thoroughly desert character, and the flora, including mountains and plains, rich in species and genera. In addition to its situation in the heart of the desert of Arizona, Tucson was centrally located, both as to position and transportation, with reference to the deserts of Texas, Chihuahua, New Mexico, California, and Sonora.

The city then had a population of about 10,000 and was

situated on the transcontinental Southern Pacific Railroad less than four days from New York, one and a quarter days from San Francisco, and seventeen hours from Los Angeles.

The Laboratory would be connected with the city by telephone, and thence in communication by telegraph and cable with the rest of the world.

The elevation of Tucson was 2,300 feet (690 m), and the highest of the mountains that surrounded the plain in which the city rested, the Santa Catalina range, reached about 6,000 feet, (1,800 m) higher. The University of Arizona, a Land Grant institution, with its School of Mines, and the Arizona Agricultural Experiment Station were also located at Tucson.

The Tucson Chamber of Commerce enthusiastically supported the new facility and provided subsidies of land for the building site; installation and construction of a water system; telephone, lights and power connection; and a road to the site of the Laboratory 2 miles (3.2 km) west of Tucson (Coville and MacDougal, 1903). Ultimately, the Chamber of Commerce obtained control of 820 acres (332 ha) which were dedicated to the use of the Laboratory.

A $4,000 ceiling for the cost of the Laboratory was set by the Carnegie Institution. Among 1903 construction costs listed were $300 for the road; $258 for the site; $926 for the power line; and an estimated $250 for the telephone line, and incidentals at $12.45.

The Laboratory (Fig. 1.1) was halfway up the mountain road to the top of the 800-foot (240 m) hill called Tumamoc, the Papago word for the horned toad (a desert lizard). The original building was L-shaped and constructed of native volcanic rock. The imported slate roof had projecting eaves for ventilation and protection from the heat. The facility included a private study, a library, a darkroom, a storeroom, and a main laboratory. In 1906, the surrounding grounds were enclosed by a wire fence of three strands on redwood posts 1 rod (5 m) apart. It was soon necessary to reinforce the fence with a basal section of close-mesh wire netting to keep out stray cows and burros and to control jackrabbits and other animals that would injure the increasing vegetation. Within a few months the difference between the vegetation on the grounds and that of the surrounding areas was noticeable, and there was also an increase in the number of roaming wild animals.

Fig. 1.1. Carnegie Desert Botanical Laboratory Building, circa 1912. (Photograph courtesy of Carnegie Institution of Washington.)

In 1906, additions to the Laboratory itself were made that extended it around two sides of a quadrangle. One side of the main laboratory was fully equipped for chemical work; the other side was devoted to work with plants, a table for microscopic work, and the thermostatic apparatus for the investigations of the effect of sunlight on respiration. Another building for chemical and physiological studies, to be called the Chemical Lab, was completed in November of that year. It was constructed of native rock with inner walls of pressed brick. It contained two laboratories equipped with water, gas, direct-current and alternating-current electricity, vacuum, air pressure, and large fume hoods. There was also a small shop and preparation room; a machine room for vacuum and pressure pump, water-still, and other equipment essential to plant physiology studies; a study, and a spacious attic. The chemical laboratory included a photochemical darkroom for work with artificial sources of light as well as sunlight, spectroscope and polariscope work. On the roof there was an isolation deck covered with sheet lead, this area served as an open laboratory for special chemical and physiological experimentation with sunlight.

The chemical laboratory burned down in November, 1938, and was not replaced by the Carnegie Institution. In addition to the valuable equipment lost in the fire, many unpublished experimental data were destroyed.

Other buildings on the grounds included a shop at the foot of the hill, constructed of the same native rock. There was also a stucco house located on land southeast of Saint Mary's Hospital. Later used as an office, the house was surrounded by several lath houses, a pump house and a small frame building affectionately called the "bug house," where insect experiments were carried out.

## The Laboratory Staff

In December, 1905, the Department of Botanical Research of the Carnegie Institution was created, and D. T. MacDougal was named director. His staff at the Desert Laboratory (see Fig. 1.2) included W. A. Cannon, B. E. Livingston, V. M. Spalding

Fig. 1.2. Desert Laboratory staff with visitors in 1906. *Standing, left to right:* Mr. Reeder, Godfrey Sykes, W. A. Cannon, B. E. Livingston, F. E. Lloyd, Mr. Lantz. *Seated, left to right:* Mrs. Sykes, Dr. Davenport, President R. S. Woodward of the Carnegie Institution of Washington, Director D. T. MacDougal, George H. Shull, Mrs. Livingston.

and Mrs. E. Spalding, and Godfrey Sykes. F. E. Lloyd was appointed for one year to complete his work upon stomata.

This eager group of investigators, having almost no previous contact with the desert, together observed and recorded a vast amount of information in an environment that must have been strange and in many ways challenging.

The early scientists, each with a decade or more of training and experience, represented a range of background and ages. Livingston, age thirty, was a plant physiologist from the University of Chicago. Cannon, in his mid-thirties, came from the New York Botanical Garden, as did MacDougal, who was in his mid-forties. Spalding, in his mid-fifties, had spent many years teaching botany and zoology at the University of Michigan.

A major influence on the Desert Lab program in the early period, MacDougal was responsible for well over one hundred publications between 1903 and 1933. His principal contribution to desert studies included work on the Salton Sea and the Colorado River Delta Region. In addition, he published studies on acclimatization, heredity, parasites, water balance, dendrographic measurements, and on long-lived cells that included desert plants. In 1920 he moved to Carmel, California, but remained in charge until 1928 when the Division of Plant Biology was established with H. A. Spoehr as chairman.

Godfrey Sykes, who had accompanied MacDougal on some of his early expeditions, was a staff member from 1906 to 1929. Much of the success of the scientific program was due to Syke's ability to keep the Laboratory operating efficiently.

Two other scientists played dominant roles in the affairs of the Desert Laboratory between 1908 and the closing in 1940. They were Forrest Shreve and Herman Augustus Spoehr.

The appointment of Forrest Shreve on May 1, 1908, was singularly important in the development of the Laboratory. Although Shreve came from the east, he rapidly became adjusted to the desert and was to become an internationally known leader in desert investigations and a preeminent figure in the new discipline of ecology.

Born in Maryland and with a Ph.D. from Johns Hopkins, he taught at Goucher College where he met and married Edith Bellamy who worked closely with him for many years at the Desert Laboratory. In addition to his laboratory and field studies, he edited the journal *Plant World* for several years until it became *Ecology*, the official publication of the Ecological Society of America of which he was a founder and later a president. Even after the Desert Laboratory closed down in 1940, the Carnegie Institution continued to support Shreve's research until he died in 1950 at age seventy-two.

Herman Augustus Spoehr accepted an appointment as a staff member of the Carnegie Institution of Washington in 1910. He was first employed to spend a small part of his time at the Desert Laboratory in Tucson and to do most of his work at the University of Chicago. It was understood that he would devote his attention chiefly to the chemical problems of plant physiology, spending only four months a year in the southwest, but within a year he moved to Tucson.

In 1920, Spoehr moved from Tucson to Carmel, California, where the Carnegie Institution's Coastal Laboratory was located. Carmel was then a small village set in the midst of a pine and oak forest, on the shores of beautiful Carmel Bay. The seacoast and the plentiful vegetation were a welcome change from the desert.

From 1910 to 1928 when he was appointed chairman of the Division of Plant Biology of the Carnegie Institution, Spoehr was largely involved with his research on carbohydrate chemistry in photosynthesis. In 1930 he spent a year as Director of the Natural Sciences Division of the Rockefeller Foundation. Returning to the Carnegie Institution in 1931, he resumed the chairmanship of the Division of Plant Biology, retiring in 1950 and dying at Palo Alto in 1954.

## The Early Years

Affairs moved rapidly at the newly established Desert Laboratory, and the research program was started before the end of 1903. Cannon was enthusiastic about the opportunities offered in this desert area. He became intrigued with the ocotillo which he regarded as a plant especially adapted to life in a desert. As he expressed it, when a stem was immersed, water "creeps up the outside at a lively rate," and he was trying to find out if it penetrated deeply. This was in November, 1903. In December, he wrote MacDougal that he was studying the role of "corrugations" of the barrel cactus and saguaro. He believed these were an essential adaptation for a plant subject to great changes in volume due to the accumulation and use of water stored in the plant.

Other scientists joined the staff during 1903 and 1904. Cannon had his studies of root systems of desert plants underway and, at the same time, was showing great interest in the water relations of desert plants. He developed a method of measuring water loss through transpiration *in situ*. He showed particular interest in the creosote bush, which is one of the most typical desert plants with ability to survive under extreme drought conditions. In March Cannon wrote to Mac-Dougal that,

*Creosote bush gives off very considerable amounts of moisture in spite of unusual dryness. With the continuation of the drought the ocotillo is dropping its leaves, and lowering its rate of transpiration of course and the bisnaga (barrel cactus) has practically ceased to give off any water.*

The scientists were especially interested in finding out what mechanisms and processes made it possible for plants to survive under very dry conditions, and they were surprised to find that desert plants often gave off "very considerable amounts of water." The solution to the problem of drought survival and the ability to live in very dry habitats was still some years in the future.

MacDougal spent many months at the Desert Laboratory in these early years planning and carrying forward investigations on soil temperature. He spent most of his time, however, in studies of the vegetation of the Salton Sea area, the delta of the Colorado River, and adjoining parts of Lower California.

Cannon's first experience with arid foresummer in Tucson in 1904 apparently was not entirely pleasurable. In a letter to MacDougal, dated June 13, 1904, he stated in part,

*Hits [sic] hot even in Arizona at 106°. I fear that Lloyd and Livingston will feel the change considerable. Tucson is getting pretty well deserted now, everyone that can raise the money has gone to the "Coast" and the rest will go in July. Even the most hardened liar among the natives will not defend this summer climate.*

His worry about Lloyd and Livingston was unfounded, as both of these men seemed to take the weather in stride.

## Early Transportation

The ability to throw a diamond hitch (used to tighten a pack rope) was the badge of the expert field traveler, as most of the earliest botanical trips were made with pack animals and would involve horseback travel for several years to come. The alpine and austral plantations in the Santa Catalina Mountains were reached only by trails, and all material had to be carried by pack. To meet this need, complete pack equipment was organized at the Desert Laboratory, including two pairs of heavy rawhide kyacks (packsacks) suitable for the transportation of instruments without damage. In addition, a fairly adequate camping outfit for mountain work was provided.

At times, in groping with the challenge of early transportation, the innovative scientists drew questions from Carnegie Institution President Woodward. A misunderstanding between President Woodward and MacDougal was expressed in a letter from MacDougal to Woodward, dated May 22, 1911, regarding the purchase of team transportation from Sykes:

*The use of the team of Mr. Sykes by the Department is by an arrangement proposed by myself in 1906, and which has been discussed in detail with everyone having an interest in the matter. The pair of draught horses belonging to Mr. Sykes are to be kept on private grounds adjacent to the Domain without expense to us. When needed, for such purpose of hauling a load or two of sand, freight up the hill, or other work for a few hours, one of our employees puts them in the harness and uses them, reporting the hours to the office, for which we pay at the rate of 40 cents per hour.*

Another item questioned by President Woodward was for the salary of a "mounted assistant," to which MacDougal replied:

*We employ Mr. B. R. Bovee as a general factotum, janitor and messenger, who with his saddle horse rides to all of our buildings and structures here early every morning, giving us a half day's service at the rate of $55.00 per month. He accompanies all mountain expeditions with his horse as packer, and when thus giving the entire day renders an account for extra services. This had proved an excellent arrangement in every [case].*

This period spanned the transition from horse to automobile transportation. The first record of the use of automobile transportation is found in correspondence for 1909 in which it is noted that a second-hand EMF was purchased in December 1908. In 1909 a new car was driven from Los Angeles to Tucson "without mishap other than a broken frame which was mended first with a mesquite pole and then a piece of gas pipe."

It was soon realized that the chief advantage of mechanical transportation for desert investigation, aside from the great increase in the radius of action, was the opportunities it gave to make dry camps when and where the researchers pleased, without being hampered by the lack of water or horse feed. The disadvantages were in mountain climbing. Slopes and precipitous hillsides, which would have been no obstacles to the sagacious horses or the condescending burros of earlier exploratory adventures, were out of the question even for the most agile of their little gas buggies. Graded mountain highways and scenic boulevards were as yet merely visionary schemes of energetic exploiters.

## The Carmel Laboratory

In 1904, the Carnegie Institution became interested in the plant breeding work of Luther Burbank, who was noted for the large number of plants he had developed and notorious for his lack of records that might serve as a guide to additional research.

Substantial grants were made to Burbank to aid his "Experiments in Plant Development." The laboratory and small field experiments of the past had never included enough species under study at the same time, and it had been impossible to draw general conclusions safely, as the numerous tribes and species of plants each had a slightly different story to relate.

By now, however, the experiments and plants included 300,000 new hybrid plums; the work of twenty-five years in crossing about 10,000 seedlings of pineapple quince; 400 new cherry seedlings; 1,000 new grapevines; 8,000 new hybrid chestnuts, crosses of American, Japanese, Chinese, and Italian species; 800 new and distinct hybrid walnuts, crosses of American black, Aieboldi, English, Manchuria, butternut, and others; many thousand apricots and plumcots; 5,000 select, improved thornless "Goumi" (Eleagnus) bushes; numerous other fruits in less numbers, and 10,000 new, rare, hybrid seedling potatoes (*Carnegie Yearbook 4*, 1905).

For eight years *Opuntia* and other cacti were secured from all parts of the world. Selections were made and crossed and thousands of hybrid seedlings raised, some tender or hardy or gigantic or dwarf, some bearing gigantic fruits in profusion and others small ones of exquisite flavor. Some large groups were developed which produced enormous quantities of nutritious food for all kinds of stock and poultry.

Dr. George Shull, at that time a young scientist who later became renowned for his work in developing hybrid corn, was sent out to assist Burbank in getting his material in shape for publication. According to W. L. Howard, (1946), one of Burbank's biographers, the relationship between Burbank and the Carnegie Institution was not a happy one for either party and Shull was the most unhappy of all. He continued, however, to collect material over the five-year period covered by Carnegie grants, although a report was never published.

Because of the difficulties between Burbank and Shull, Cannon at the Desert Laboratory was assigned to leave Tucson and work with Burbank. He spent the summer months of 1907 and 1908 at Carmel-by-the-Sea, not far from Burbank's experimental garden where he came in contact with the Carmel Development Company, whose executive officers saw the wisdom of offering to build a laboratory in Carmel-by-the-Sea.

This suggestion received the favorable attention of President R. S. Woodward, who wrote a letter on December 21, 1908, to the Carmel Development Company accepting their offer. The Carmel Laboratory became the Desert Laboratory summer headquarters in 1909.

During this same period, and the subsequent years, it appeared that the honeymoon with the Chamber of Commerce at Tucson was over. In a letter from MacDougal to the Secretary of the Chamber of Commerce, dated February 16, 1910, he noted:

*that the telephone has been removed from the Desert Laboratory in accordance with the procedure described by you. The failure of the Chamber of Commerce to make good its agreement to maintain a telephone at the Desert Laboratory throws some additional expense on us for messenger service, part of which will be personal. I am therefore compelled to withhold my subscription to the Chamber of Commerce in order to meet this.* . . .

By 1913 the Laboratory at Carmel had been established as an integral part of the Department of Botanical Research as shown on the letterhead which carried at the left: Coastal Laboratory, Carmel, California, and at the right, Desert Laboratory, Tucson, Arizona. By 1920, the year that Spoehr moved to Carmel, MacDougal stated in a letter to Merriam, at that time President of the Carnegie Institution of Washington, June 12:

*The Tucson establishment is practically closed this summer as all of the staff are either here or in the field.* . . .

In 1921, a letter from MacDougal to the Carmel Development Company indicated that he had about completed construction of a house at the "highlands." After that he spent very little time in Tucson. Shreve, however, became even more firmly rooted in Tucson with the passing years. He also was a fixture at the Desert Laboratory and remained a staff member until his retirement in 1948.

## Desert Laboratory, 1910–25

The years 1910 through 1925 were spent making notable advances in soil-plant relationships, plant physiology, especially moisture relationships and ecological studies of individual species. Cannon made his comprehensive studies on the root habits of desert plants, and traveled to Algeria, Australia and southern Africa to study these deserts as contrasted with those of the southwestern United States. Livingston conducted diversified studies on soil and plant moisture relation. Mac-Dougal was interested in genetics, induced parasitism, water relations, continuation of studies on the Colorado Delta and the Salton Sea, plant acclimatization and, during the latter part of the period, growth relationships and processes of woody plants. Edith Shreve carried out extensive research on transpiration and autonomic movements in succulent plants. Forrest Shreve studied the establishment and survival of succulent and woody plants, the effect of low temperatures on plant distribution, the vegetation of the Santa Catalina Mountains, and with Livingston, distribution of vegetation in the United States—physical factors determined by slope and exposure—and toward the end of this period began publishing articles on some North American Deserts. Spoehr carried out extensive studies on photosynthesis, respiration, and metabolism. In addition, a variety of papers was published on desert-related subjects by short-time laboratory associates.

While these various studies were under way, changes were being made in the organization of biological investigations in the Carnegie Institution of Washington. The first of these involved changing the name to Division of Plant Biology.

Dr. Spoehr was appointed Chairman of the Division of Plant Biology with the understanding that he would have general supervision of the program for unification of the interests of the Institution in plant biology.

The reorganization and general program was described in the President's report in *Carnegie Yearbook 27*:

*Following consideration of possible means of bettering organization of the Institution in the field of plant sciences, the Trustees authorized a re-statement of our program beginning with the year 1928. The many important activities in this subject have been brought together as a Division of Plant Biology under chairmanship of Dr. H. A. Spoehr. It is not the intention through setting up the present organization to extend the work of the Institution so as to cover the whole of plant biology. It is rather the purpose to obtain a somewhat better visualization of the field, and of the opportunities for effective cooperation among the interests concerned. The grouping is designed to give opportunity for maximum unity in operation of the diverse interests involved, and at the same time to increase the possibilities of effective work for each group.*

*The division is composed of the following sections: Photosynthesis, Physiology and Plant Growth, Desert Laboratory Studies, Plant Ecology, Taxonomic Relations and Paleobotany. Each section is under guidance of a leading investigator. Through cooperation of Stanford University a lease for five acres of land has been obtained on the campus of the university. Provision has been made for a new laboratory to serve as headquarters of the Division, and for conduct of researches in photosynthesis, taxonomic, and ecologic problems, and other subjects.*

*The groups of researchers within the Division will operate at the localities at which the work can be conducted to best advantage. So, the Desert Laboratory at Tucson, Arizona, in immediate charge of Dr. Forrest Shreve, will continue as the point at which investigations, most effectively conducted under arid conditions can be carried on with advantageous laboratory facilities, and with utilization of a diversified "campus" covering approximately 800 acres of extremely interesting desert country remaining completely protected under primitive conditions.*

At the laboratories at Carmel, California, the program of researches on the physiology of tree growth continued under direction of D. T. MacDougal, who had recently had associated with him two distinguished investigators, J. B. Overton, of the University of Wisconsin, and Gilbert Smith, of Stanford University. Their series of studies made significant contribution toward interpretation of the structure and physiological process of the tree, coming at a critically important time, when intensive investigations of such problems were of large importance in the general field of forest research.

The period between January 1, 1928, and the closing of the Desert Laboratory in 1940 was marked by the leadership of Forrest Shreve. Although his ties were maintained with the divisional leadership in California, the Laboratory was largely independent.

Activities during this period included laboratory research, various plant physiological studies conducted on the laboratory grounds, and soil and climatic investigations, in addition to desert explorations mostly centered within the Sonoran Desert.

The technical staff during this period consisted of Forrest Shreve and T. Dwight Mallery, principal assistant, together with younger assistants on assigned tasks. These included W. A. Turnage, A. L. Hinckley, and R. R. Humphrey. In addition, several visiting scientists were in residence for periods varying from a week to a year. Howard Gentry, although not a member of the laboratory staff, was provided space and facilities while working on a Rio Mayo flora project. The laboratory facilities and grounds were also used cooperatively by scientists from the University of Arizona and other institutions.

## Closing of the Desert Laboratory

In the late 1930s it became evident that the research program financed by the Carnegie Institution would have to be reduced because of decreased income from investments.

According to diary information supplied by Margaret Shreve Conn, the first intimation of the possible closing of the laboratory came in 1938. This was documented by an entry in Shreve's diary, "Mr. Varela arrives. . . . He seems to be cheerful about everything except the financial future of the Institution."

According to Mrs. Conn: "All through 1938 and 1939 there are references to people visiting 'the hill,' as Dad always referred to it, and of his going to 'the hill.' " The first actual reference in his diary was July 7, 1939; "In Palo Alto. Go up to see Spoehr and discuss the closing of the laboratory."

In 1939 negotiations with the University of Arizona to take over the Laboratory were unsuccessful, and the buildings and grounds were later offered to the U.S. Forest Service to house the offices of the Southwestern Forest and Range Experiment Station. The deed to the U.S. Government is dated August 7, 1940.

The decision to transfer the Desert Laboratory to the Forest Service was reported in the President's report in *Carnegie Yearbook 39:*

*In terminating our effort at the Desert Laboratory, at Tucson, Arizona, we have succeeded in avoiding the sacrifice of valuable elements. The work of Dr. Shreve will continue under our auspices. Much of the program which the Institution there initiated many years ago has become incorporated into the large operations of the government in connection with forestry, soil conservation, and the public domain. We have therefore transferred the property at Tucson to the Forest Service for use in its research program, which has objectives closely parallel to those which prompted the early establishment of this laboratory. In this connection they plan to maintain the isolated area which has been preserved by the Institution for many years.*

At the time the Forest Service took over the Laboratory, Arthur Upson, director of the Southwestern Forest and Range Experiment Station, was sympathetic toward desert investigations. The property was later made a part of the Coronado

National Forest, and a less sympathetic Regional Forester allowed many easements in the fenced area.

The property acquired by the Forest Service was put up for sale to the University of Arizona in 1960.

As mentioned previously, the Desert Laboratory and its facilities had been offered to the University of Arizona before they were offered to the Forest Service. The University of Arizona could have had the whole package for one dollar. In 1960 it paid the Forest Service $140,500 for the buildings and grounds.

Although activities had slowed down considerably for the last twelve years before the Laboratory was closed, the impression lingered that this was an institution devoted to high quality research and that the scientists were dedicated to this research. They believed that the acccumulation of knowledge related to deserts ranked high among scientific endeavors.

# The Nature of Deserts

## What Is a Desert?

TO BE STUDIED, deserts must first be identified and their location on the globe must be known. Inquiry into the activities of the Desert Laboratory reasonably begins with the question "What is a desert?"

In their attempt to answer this question, Coville and MacDougal found that in the first half of the nineteenth century, as shown in textbooks as late as 1843, the vast region between the Missouri River and the Rocky Mountains was designated as the Great American Desert, largely because of a lack of definite knowledge by early geographers. Such designation was shown in early maps, circa 1835–45.

*The insufficiency of [this] description obviously rests upon faulty observation, and upon the failure to recognize the fact that the habitability of a region is no criterion of its arid character. The development of modern methods of transportation has made possible the maintenance of dwellings and towns with considerable population at 100 or 200 miles (160 or 320 km.) from the nearest supply of water. [Coville and MacDougal, 1903]*

The authors noted later that, based on the results of more recent explorations and surveys, the arid regions were fairly

17

well delimited and the southwestern desert areas—the badlands, the staked plains of Texas, the Chihuahua Desert, the Great Basin, and Colorado Desert—were shown approximately within the region which was appropriately designated as desert.

According to conceptions at that time, deserts were rainless, usually sandy, and commonly not habitable. They might be regions of considerable extent that were almost but not quite destitute of vegetation and hence uninhabited chiefly on account of an insufficient supply of rain (such as the Sahara and the Asiatic deserts.)

The term desert was chiefly and almost exclusively used in reference to certain regions in Arabia, northern Africa, and other lands in Central Asia. The only region in North America to which the word applied was the Great American Desert, a tract of country south and west of the region of the Great Salt Lake.

Coville and MacDougal believed that desert conditions could exist when evaporation was greater than precipitation and that extreme desert conditions could occur in some portions of the tropics where high evaporation occurred along with rainfall of up to 70 inches (175 cm) per year.

They also observed that regions in which precipitation is less than evaporation are characterized by a lack of running streams or permanent runoff. However, the rainfall in the desert may be so heavy in certain seasons as to produce floods that, rushing downward over the slopes and mountainsides, produce distinct streamways extending for miles on the plains below. These flows soon cease after the rains have passed and the streambeds become dusty empty channels until the next rainfall. The investigators considered that seasonal distribution of rain was very important. If the rainfall occurred within a brief period, especially during the months of low temperature, the resulting dryness of the growing season would result in desert conditions.

In a later publication, MacDougal (1913e) called attention to the geological features of deserts, noting that the vegetation of arid regions is striking to the casual observer, but it is no less characteristic than the surface geological features. Rocks crumble and the fragments fall to the bottom of the cliff and roll

down the slopes everywhere, but the character of the disintegration of the mountain walls and the position that the detrital
material takes at their bases is largely determined by the
amount of precipitation and is different in areas where precipitation overbalances evaporation. The masses of detritus that
are carried out of the mountain canyons run out as rounded
tongues of loose material for a distance of a few to 15 or 20
miles (24 to 32 km) and, by reason of their arched or curved
surfaces, are known by the Mexican term *bajadas*.

Further, while the lack of water is the chief factor in the
production of deserts, they may be produced as a result of
defective nutritive or mechanical conditions. Such conditions
are to be found in areas in which the soil contains harmful
substances in injurious concentration, of which the alkalies are
familiar examples. Evidence of unsuitable mechanical conditions are offered by sand dunes in many parts of the world. In
sand dunes the substratum is in constant motion of greater or
less rapidity, may lack suitable water supply, and be devoid of
other nutritive material. Even though the dune areas may be
supplied with water in proper quantities, the character and
movements of the substratum result in unfavorable growth
conditions.

In his description of the botanical features of the Algerian
Sahara, Cannon (1913a) observed that environmental conditions encountered by plants in the arid regions are widely
different from those of moister regions. Precipitation is not
only slight, but it shows an enormous range in variation from
year to year. The rate of evaporation is high, the temperature
of air and soil varies widely both during the day and with the
seasons, the light is of great intensity, and the soil is low in
humus content and may contain an excess of salts. These most
striking physical factors of deserts are present in combinations
that result not only in differences among deserts in arid regions
but may include much variation within their borders.

Shreve (1925) observed that the biota of the deserts can
be distinguished from that of other regions by recognizing a
large group of characteristics, any one of which may be found
in plants of other environments. Among these, the most common are low stature, the smaller number of individuals per unit
area, a deep-seated and widespread root system, and structural

and physiological capacities for maintaining the balance between the income and outgo of water. Among desert animals
there are structural features that appear to aid in the conservation of water and make subsistence possible without liquid
water through use of moist food or the metabolic water of dry
food. Nocturnal subterranean habits enable these animals to
evade the highest temperatures and to reduce the expenditure
of energy and water that would result from exposure to the
high evaporating power of the ground at midday.

In this same article, "Ecological Aspects of the Deserts of
California," Shreve (1925) defined the desert as follows:

*It is impossible to define desert in terms of a single characteristic, just as truly
as it is impossible to differentiate species by such a procedure. A low and
irregularly distributed rainfall is the fundamental feature but is not to be
taken as the sole criterion of the environmental conditions. With low rainfall
are associated low moisture content of the soil and low relative atmospheric
humidity, often fluctuating only by reason of the daily changes of temperature.
The diurnal temperatures are high, but the daily range is great, owing to
the active nocturnal radiation from a bare and dry soil. The air movements
are greater than in a humid climate, being manifested both by wind and by
small cyclonic disturbances, or tornillos. The evaporating power of the air
is great by reason of the combined effects of low humidity, high temperature
and great air movement. . . . Low precipitation and torrential rains are
responsible for poorly developed drainage, the occurrence of dry lake beds, the
existence of steep outwash slopes, stretches of sand constantly shifted by the
wind, soil highly impregnated with salts, surface mulches of small stones
covering a fine soil, and many other minor features which are of great
importance in determining the local distribution of plants and animals.*

## Evolution of Deserts

The enigma of the desert is its life. Our notions of living
things, their form, how they function, and their relations to
each other, were long formulated from experience and observation in temperate or tropical regions. The contrasts the early
scientists first encountered on the desert seemed startling and
harsh. The true depth of these contrasts and their objective
significance reveal themselves only gradually through intimate
living in the environment. Whence came these desert plants
and animals and how are they able to maintain their life, to
adjust their functioning to the extreme physical conditions to

which they are exposed? The plant organisms function as do all of the higher plants—in manufacture of their own food, growth, and reproduction—an astounding capacity to modify selectively the ways of their ancestors and to become organized for the environment in which they find themselves (Shreve, 1925).

The physical environment of the desert alone presents an extremely complex picture, especially when regarded in its relation to the plant life of the region. Not only do the various climatic factors (such as rainfall, evaporation, and temperature) vary over a wide range, from extremely unfavorable to merely difficult, but the year-long distribution of such climatic factors is equally important for survival. The long and often irregular periods during which the plants are exposed to extremes impose rigors which are definitely determinative for survival (Spoehr, 1956).

A single season or year, or even a few years, of observation yields only a fragmentary picture, and often an erroneous one, of the interplay of climatic factors on living organisms. The variations in climate and whether they are truly cyclical or only approach cycles in the astronomical sense, as well as the definite correlations between solar and terrestrial phenomena, obviously require extended study and elaboration. It is clearly established that there are recurrences of climatic conditions favorable and unfavorable to the development of plants. Moreover, it is known that the great mass of observations on the relationships of plant and animal life to these environmental conditions is definitely amenable to scientific collation. The concepts arising from such investigations ultimately have real social significance. They constitute a basis for solution of the problems which even today are presented in the management of the half-billion acres (200,000,000 ha) of arid and semi-arid range land in the United States, and the larger areas in Africa, Australia, and South America.

## Origin of Desert Plants

The question of where desert plants came from was considered by several of the scientists at the Desert Laboratory. MacDougal (1909c) began with the origin of life, with the early plants largely restricted to water—or at least moist—habitats.

He followed step-by-step the development of desert vegetation. He believed there was undisputed evidence that extensive arid areas existed in all the great geological periods. He pointed out, however, that much of the origin of desert plants is open to speculation because formations that gave evidence of desert conditions are notably free from fossil plants and contain but little in the way of animal remains.

MacDougal thought that predesert forms may have been able to maintain themselves on mountain slopes within the desert area where precipitation was higher and the evaporating power of the air was lower. Plants were able to invade the desert during seasons of more favorable conditions. For example, annual plants, even with a mesophytic habit, might develop from sprouting seed, carry through their cycle activity, and remain dormant in the form of heavily coated seeds during the warmer, drier part of the year. Perennials with deciduous leaves also might meet this seasonal situation. MacDougal pointed out that many of the annuals occurring in the Sonoran Desert are in one sense not xerophytes since they require as much moisture for their development as plants in Maryland, Michigan, or Florida.

Southwestern North America has been arid for an extremely long period but not uniformly so, and variations in climate with regard to temperature and moisture that could be of profound importance have taken place within the last 2,000 years. It seemed fair to assume that similar oscillations have taken place previously, each movement extending perhaps over a few hundred years (MacDougal, 1911i).

MacDougal concluded that the desert conditions encountered on the earth's surface at the present time are not to be taken as having come about by a simple, direct, and continuously increasing aridity. He believed evidence from Asia, Australia, and northern and southern Africa, and the plains of western America established that climates have undergone oscillations between periods of lower temperature and maximum precipitation and of maximum aridity with increased variations in temperature; the swing from one maximum to another occupying periods of a thousand, two thousand, or many thousands of years. Within these fluctuations there may have been minor fluctuations of a shorter duration.

It is under these conditions that MacDougal (1911i) thought the evolution of the desert vegetation of the Southwest took place. Available knowledge compelled him to believe that much of it originated somewhere within the limits of the region that was still arid.

He pointed out a number of forms in every desert with structural features that give them a distinct aspect and with many physiological capacities definitely suitable for activity during the drier seasons and that may remain inactive during other seasons. Perhaps the most important constituents of this specialized flora are the cacti that must have originated somewhere in the Mexican highlands during the Tertiary period or later. This group extends through South America and contains more than a thousand species with distribution offering some most highly localized occurrences of species.

All attempts to find paleobotanical records giving evidence that desert plants lived in previous arid climates had so far failed. But MacDougal believed consideration of the facts led to the inevitable conclusion that the form-characters, moisture-conserving capacities, and resistance to desiccation distinctive of xerophytic species must have made their appearance within comparatively recent geologic time.

The development of the desert biota was found to constitute a problem that is largely distinct for each of the great continental desert areas—that there are very few species of plants or animals common to the arid lands of America, Africa, Asia, and Australia. Certain relationships could be cited between the deserts of North and South America and between the desert areas and the humid regions that are nearest them. These are much greater than the deserts that are far apart. In other words, deserts have been populated by access of plants and animals from near at hand, and only to a very slight extent by races from other deserts.

Shreve (1936a) stated the plants that are confined or nearly confined to the desert are living representatives of races of plants that have undergone the greatest change in entering the desert and have acquired all the characteristics that have made possible their survival and success. In the evolutionary development that has accompanied entry into regions of aridity and brilliant sunshine, features such as flowers, fruits, and seeds have been little affected. In a greater number of cases, the plants have undergone modification of the root, stem, leaf, and other vegetative structures. Some of these changes have been so profound as to make the appearance of the plant wholly unlike that of its nearest relative, yet the flowers and fruits are almost unchanged.

## Types of Desert Plants

The succulent group of plants was seen to be chiefly represented by the cacti in America and the euphorbias in South Africa. Succulents are rare in other deserts; however, cacti introduced from America into Northern Africa and Australia have flourished. In the stem succulents, the leaf has been dispensed with and its work has been taken over by the green stem tissues. The stem has undergone enlargement through the development of great masses of tissue in which an accumulation of water is held. The root system of the succulent is widely distributed near the surface of the ground and when rain wets the uppermost limits of the soil, the plant quickly renews its water content.

The water supply of nonsucculent perennial plants must be renewed from time to time throughout the frostless season. Their survival depends upon a root system, widely distributed in the soil in which some moisture may persist throughout the year.

Shreve summarized his observation by stating that the ephemeral plants escape the rigors of the dry climate, the succulents have a mechanism which equalizes the irregularity of water supply, and the nonsucculents have the daily problem of maintaining equality between their water loss and water supply. These three types of adjustments are found in hundreds of species and the three types may grow side-by-side in the desert vegetation.

Succulent plants are a well developed group with a very important place in the vegetation of arid America. Only a few families have developed succulence in America, and the cacti are by far the most important of them with over 1,200 species. This great group originated somewhere in tropical America and made its way from the forests through the semiarid caatinga and thorn forest into the deserts north and south of its center of origin.

Another group adding variety to the vegetation of American deserts are the plants in which the leaves as well as the stems are succulent. The largest of the leaf succulents are the century plants, in which the stem is thick and succulent but so greatly shortened that the leaves arise very close together. Smaller plants, similar to the century plants, include the dudleyas, which are particularly abundant in Baja California.

In comparing the Sonoran Desert with the Karroo Desert in South Africa (with respect to development of the succulent habitat) Shreve pointed out that there are no cacti, no century plants, and no dudleyas, and yet there are plants closely resembling each of these. Similar conditions have brought about very similar types of plants, but they have little or no family relationship. In the Karroo, the place of the cactus in the landscape is taken by members of the Euphorbia family, the place of the century plant is taken by the aloes, and the dudleyas by hawarthias and gasterias. In outward appearance these related plants have made closely similar development in the course of their adjustment to nearly identical climates. Furthermore, there are a large number of nonsucculent plants in which there is greater variety of structure than in the succulents. There are also differences in the vegetative processes of the plants and their adjustments to environment, and especially great differences in the duration of the life of the individual plants in their separate branches. In the woody forms, the stem may have the normal type of structure found in hardwood trees or may depart from it in nearly every feature. The surface of the stem may bear a rough bark or may be smooth and green, carrying on the principal functions of the leaf.

Shreve (1942a) pointed out that the most universal feature of specialization in desert plants is small leaf size. This is conspicuous in the members of a wide range of families and is true of the plants in all but the most favorable habitats. The prevailing leaf area is less than 0.15 square inch (1 square cm) in area. The reduction of leaf size and the shortness of the period in which they can be active have greatly impaired the essential function of the leaf in desert plants. In a large number of perennials the stem has assumed these functions and is provided with stomata and chlorophyll-bearing tissue, which in some cases continue to function on large tree trunks over two hundred years old. The leaves of the green-stemmed trees and shrubs may be small and short-lived or may be wholly absent in the adult plant.

The leaves of nonsucculent plants may be perennial or may appear only in the warm season, or only in the wet season, or may even require a season that is both wet and warm. The size of leaves is varied, but small ones predominate. In some, such as the richly branched foothill paloverde, there are minute leaves in the rainy season, but their total area is much less than that of the green twigs and stems that are doing most of the photosynthetic work. Leaves are found on very young smoke trees, but they are minute or absent on old ones, and the crown is made up entirely of richly branched twigs. Allthorn seedlings have leaves for a few weeks, but the mature tree is green-stemmed and leafless.

As a part of the desert aspect Shreve mentioned that trees with dead branches and twigs are conspicuous. These represent the death of plant parts due to exceptionally dry periods. In effect, the twigs and branches are deciduous, much as the leaves are in hardwood trees. In prolonged rainless periods, there may be a high mortality of leaves and branches but rarely are the crowns of their root systems destroyed. Surprisingly, some ferns are true and successful desert plants. Their drought-resistant protoplasm endures the unfavorable seasons without any of the elaborate mechanisms of the larger plants.

Shreve believed that the nonsucculents have displayed different modes of adjustment to desert conditions, achieving different degrees of success. The criteria for success require that a plant must have a large area of distribution but be abundant in some part of its area; it must show some degree of elasticity in its habitat requirements and must have solved the problem of withstanding the longest dry periods to which it is subject by normal climatic fluctuations. Judged by these standards, not more than 20 percent of the nonsucculent species have achieved a high degree of success.

The factors that are possibly responsible for the development of characteristics of desert species may be due in part to the rigors of the environment, but they also are indirectly subject to the depredations of various animals. MacDougal (1908g) believed that if it were possible to exclude certain birds and mammals from the domain of the Desert Laboratory, the included square mile (2.6 km) would soon become a dense forest of saguaro. In addition to the enormous consumptions of seeds, almost all of the seedlings that have a swollen hypocotyl are appropriated by animals since these present about the only available supply of water in the dry aftersummer during which they come into activity. "The actual competition is not between the saguaro and other plants, but between the great cactus and certain animals, with the water supply as the indirect crucial factor" (MacDougal, 1908g).

## Deserts of North America

The descriptions of the deserts of North America are largely taken from Shreve's publication, "The Desert Vegetation of North America" (Shreve 1942a). I have also made extensive use of the writings of other authors and of my own experience and knowledge.

The area occupied by the North American deserts extends southward from central and eastern Oregon, embracing nearly all of Utah and Nevada, southwestern Wyoming except the higher mountains, and reaching westward into southern California to the eastern base of the Sierra Nevada, San Bernardino, and Cuyamaca mountains. From southern Utah the desert extends into Arizona and the lowland of Sonora, as far south as the delta of the Yaqui River, and south along the coast to northern Baja California, in the lee of the Sierra Juarez and Sierra San Pedro Martir. South of these ranges it extends across the peninsula as far south as the northern end of the Sierra Giganta. From there it is limited to the Pacific Coast, a very narrow strip along the gulf coast, and parts of the lower elevation of the Cape region.

In the highlands of southeastern Arizona and southern New Mexico, the continuity of the desert is broken by desert grassland transition. It appears at a lower elevation in the valley of the Río Grande and the Pecos River, extending as far east in Texas as the lower course of the Devils River. In Mexico the

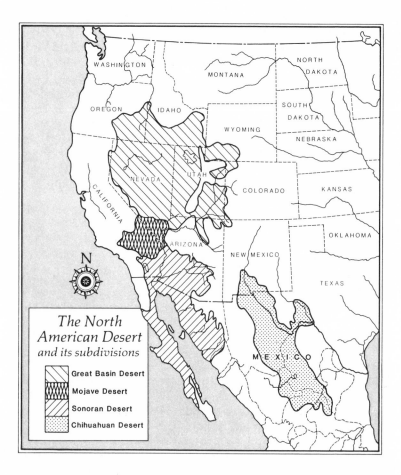

The North American Desert and its subdivisions

Great Basin Desert
Mojave Desert
Sonoran Desert
Chihuahuan Desert

desert extends continuously through eastern Chihuahua and nearly all of Coahuila, being broken by only a few higher mountains and elevated areas of grassland. Farther south the desert is confined to eastern Durango, northern Zacatecas, the western margin of Nuevo León, and the northern part of San Luis Potosí. There is also an isolated area of desert in the Columbia River basin in eastern Washington, while in the south there are several detached areas which occur in Hidalgo and Puebla, notably the valleys of Ixmiquilpan, Actopan, Mesquital, and Tehuacán. Around its periphery, the desert merges into other types of vegetation—though each of these plant formations exhibits features of similarity to the nearest parts of the desert, thereby increasing the difficulty of placing sharp boundaries on the region.

The stability of continents and the character of air movements over a long period give good ground for postulating the continuous existence of large desert areas in North America since the early Cenozoic. These areas have undoubtedly changed in size, shifted in position, and fluctuated climatically, but the present plant population of the desert is a result of these changes operating on the plants of the desert itself and on those constantly entering it—and at all times, they enter in a favorable period.

The northern Great Basin is the coldest part of the North American Desert, and, although it has large numbers of species, it is poor in life forms and its communities are simple in composition. This is also true in much of the Mojave Desert and the very arid parts of the Sonoran Desert lying near the head of the Gulf of California. Throughout these areas, variety in vegetation is due to changing temperature resulting from differences in altitude and latitude and to some of the larger soil differences. In the colder and drier parts of the North American desert uniform plant communities commonly extend over large areas, covering hills and plains down to the very edge of the streamways. In the driest parts of the Chihuahua desert there is a similar, but less marked, lack of habitat differentiation in the vegetation. Areas in southern Arizona, Baja California, Sonora, Chihuahua, Durango, and Zacatecas have mild winters and an annual precipitation exceeding eight inches (200 mm). In this region there is a strong habitat differentiation of the vegetation closely controlled by the physiographic pattern and by accompanying differences in the moisture of the soil.

Although nearly one-third of the flora in the North American deserts is composed of members of the legume family, the sunflower family, and the grass family, there are many other families that are strongly represented. The total number of families of seed plants is 137. In comparison, there are 17 families in the northeastern United States. The predominant vegetation is shrubbery. Desert shrubs fall into two rather well-defined groups: one is characterized by moderately branched, woody stems that undergo a definite seasonal termination of growth. The creosote bush and catclaw are typical examples.

The members of the other group have soft or brittle stems, are very profusely branched, and have indeterminate growth without formation of resting buds. Examples of these are sagebrush, rabbit brush, saltbush, bursage, and brittle bush. These semishrubs decrease in dominance from the north and the highlands toward the south and the lowland. The impression of monotony given by the vegetation of large sections of the desert is due to the shrubs, one of which is usually dominant, or by a dominance of several shrubs with similar life form, height, and color. In the northern, more elevated parts of the desert, the precise conditions responsible for determining the boundaries on all sides of the desert are complex. Livingston and Shreve (1921) placed the limit of the desert in the United States as the isoclimatic line along which the ratio of rainfall to evaporation is 0.20. Shreve (1940) pointed out that biological phenomena rarely admit to explanation on the basis of a single determining factor, a single ratio, or other composite expression of a small group of factors. The edge of the desert is the approximate line of demarcation between the distributional areas of two sets of plants with different life requirements. Not only do these requirements differ for the various plants in one section of the edge, but they differ still more for the dominant plants in the several sections of the edge. For example, the chaparral of California and Baja California differs from the nearest desert in the greater density of chaparral, in its predominantly evergreen sclerophyllous foliage and its strong uniformity of life forms. The pinyon-juniper woodlands that border much of the northern desert at the same time have much of the openness of the desert and some of the desert's diversity of life forms. The grassland that borders the desert of New Mexico, Texas, and Chihuahua has a much heavier carpet of grasses than all but exceptional local habitats of the deserts and is very poor in shrubs, succulents, and small trees. The arid brushland along the eastern edge of the desert in Texas, Coahuila, and Nuevo León is richer in shrubs and small trees than the adjacent desert. The thorn forest that lies immediately south of the desert in Baja California and Sonora is a closed or nearly closed stand of drought deciduous and winter deciduous trees with some evergreen trees, large succulents, and many shrubs. It has little in common with the desert except for the variety of life forms and the presence of large and small stemmed succulents.

Certain parts of the desert edge are sharply drawn, especially where high mountains cause a marked change of environment within a short distance. Other portions are broad and are occupied by transitional belts in which slope exposure, character of soil, and other local conditions may favor one or the other of the two adjacent vegetations.

The climate that determines surrounding vegetation types is related to climatic conditions of adjacent desert sections. The physiognomy and the structure of the desert communities have some relation to those of the adjacent vegetations, and the complexion of the desert flora is invariably colored by nearby floras.

In some cases plants common to the desert extend beyond its areas or habitats. An example is big sagebrush, one of the leading dominants of the Great Basin, that ranges into the coniferous woodland and into open coniferous forest around the Great Basin.

Common plants surrounding the desert behave in much the same manner as the desert plants do. They may not approach the edge of the desert or they may reach it in almost undiminished numbers and enter it for a short distance in favorable habitats. In many places in Nevada, Utah, and Arizona the juniper enters the desert for short distances, and its presence tends to obscure the true edge of the desert.

The extensive grassland in western Texas and New Mexico is a transition region where the conditions are intermediate between the optimum ones for the grassland and for desert. The vegetation is formed by an infiltration of plants from each of these areas, with very few dominant species that are distinctive of the transition region. In northern Chihuahua there are a number of *llanos,* or grass-covered plains, that are not related to the true grassland and its transitions but are distinctly a desert association occurring also in Arizona and Sonora.

The northern plateau has been mapped as an extensive undrained basin called the *Bolson de Mapimi.* But, in fact, it includes a large number of independent undrained basins, varying from a few hundred to several thousand acres (100 to 1500 ha) in size. In such llanos there is usually an almost pure stand of tobosa grass or halophytic grasses. These llanos are not regarded as part of the climatic grassland formation but rather as a desert association controlled by soil conditions. Surrounding and overlooking the llanos are invariably long outwash slopes covered with typical desert of creosote bush and tar bush.

Along the southwestern margin of the Chihuahuan desert there is a distinctive type of vegetation between the desert and the grassland. This is essentially a thin cover of short grass with a continuous open stand of small trees and tall prickly pears. Shreve did not feel that this cactus-acacia-grassland should be regarded as a transition between desert and grassland. In physiognomy it carries no suggestion of either. It has most of the typical grasses of the grassland, but its flora includes none of the characteristic plants of the desert except for occasional mesquite trees (Shreve 1942b).

Shreve divided the desert area on the basis of community structure and floristic composition. Substantial differences in latitude, altitude, nearness to the sea, and difference in topography and minerology sometimes affect large areas. The subdivisions recognized were the Great Basin Desert, Mojave Desert, Sonoran Desert, and Chihuahuan Desert.

Geographical, rather than vegetational, designations were used for subdividing because they are simpler and more generally understood as they have long been in use among plant and animal geographers. Some botanists have called the part of the Sonoran Desert that lies in California by the name Colorado Desert as applied by Blake in 1853 in reference to the Colorado River before the name was applied to the state of Colorado. This is a rather ambiguous name, as it overlooks the fact that the area in question has no features that distinguish it from the adjacent parts of Arizona, Baja California, and Sonora.

These deserts with the exception of the Sonoran Desert, to be covered separately, are briefly discussed below (Shreve 1942a).

## The Great Basin Desert

According to Shreve (1942a) the Great Basin Desert is nearly coextensive with the Great Basin Province with extensions into adjacent parts of the Colorado Plateau. It includes the Harney Basin in Oregon and the Snake River Plains in Idaho and extends south through southern Nevada and Utah reaching eastward into the Red Desert of southwestern Wyoming, the western border of Colorado and the Northeastern corner of Arizona.

The desert lies largely above the 4,000 feet (1,200 m) of elevation with a rainfall of four to eight inches (10 to 20 cm), increasing to 11 to 13 inches (28 to 33 cm) at higher elevations and in the northern extension. The precipitation is more uniformly distributed through the year than in other desert areas. But in most localities the rainfall is heaviest in the spring months with June being the driest month. The rains of midsummer are light and there is no protracted dry season. In conjunction with the moderate temperatures that result from higher latitude and altitude, these factors give the Great Basin more favorable moisture conditions than the low precipitation would indicate. The areas are subject to frequent periods of freezing temperatures—for as much as a week during the winter months.

The Great Basin consists of a large number of undrained minor basins separated by fault block ranges or by great accumulations of outwash material. Usually at the center of each basin there is an alkaline flat or dry lake. Around these are concentric belts in which the soil conditions gradually change outward toward a coarser texture and a lower salt content.

The salient features of the vegetation of the Great Basin Desert are a simplistic composition and fidelity and distribution with relation to belts of soil conditions. In the valleys and on outwash slopes there are nearly pure stands of semishrubs that often extend for 20 to 60 miles (32 to 96 km) with little change. The rather monotonous community is interrupted only by rocks, much sand, a broken irregular surface or alkaline flats or dry lakes. The most common plants are big sagebrush and shadscale. The two are often associated, but the shadscale appears more often in a lower belt than the big sagebrush and in more alkaline soils.

Sagebrush is dominant throughout most of the area extending into the woodland and pine forests as well as into the Mojave Desert. It shows variation in density, height, rate of growth, and size of individual plants under different habitat conditions. The height is commonly 2 to 4 feet (0.6 to 1.2 m) and seldom exceeds 6 or 7 feet (2.0 or 2.1 m). The roots often penetrate to a depth of 10 feet (3 m).

There are, according to Shreve, 19 species of sagebrush and 14 species of shadscale found in more or less close association within the Great Basin and accompanied by other native perennial species. In addition to these leading plants there are a number of semishrubs important in the vegetation of the Great Basin Desert, either in pure stands over large areas or as dominants in association with sagebrush and shadscale. These include species of rabbit bush, hopsage, winter fat, several species of saltbush, and in the southern part of the desert, blackbrush, which is found on coarse soils, very low in salt content. Blackbrush forms rather extensive pure stands and in many places marks the transition between the Great Basin Desert and the Sonoran Desert.

Grasses and sedges are not important except for the sod of saltgrass and clumps of alkali sacaton in highly saline or alkaline flats. At lower elevations grass is rarely seen, but occasionally stands of galleta are evident. In Nevada, north of the Humboldt River, grasses play a more important role. Hills and slopes are often clad with grasses rather than shrubs. These include species of bluegrass, wheatgrass, and festuca, and the slopes usually mark the limit of the desert.

## The Mojave Desert

This smallest subdivision of the North American Desert lies almost wholly in California. The name is used often in a restricted sense, applying only to the southwestern corner of the area. The exploration carried on by the Death Valley Expedition and reported by Coville (1893) was almost entirely in the Mojave Desert. The boundaries of the Mojave Desert extend east and north from the San Gabriel and San Bernardino mountains, lying between Death Valley on the north and the eastern end of the San Bernardino range on the south, having no definite eastern limit, and including in the latter an indefinitely outlined area adjacent to the Colorado River (Shreve 1925).

The Mojave Desert is an integral and continuous portion of the great arid region of the southwestern United States and Mexico, and some of its features are to be best understood when considered in connection with adjacent areas. The Mojave Desert differs from the adjacent desert areas in the occurrence of the principal part of its rainfall in late winter. Vegetation is rich only where the late winter rain is heavy enough to assure a relatively moist soil for several weeks after the temperatures have begun to be favorable for growth.

The portion lying nearest to the fringing mountain ranges possesses a vegetation in which there are more large plants, a greater number of plant individuals per unit area, a greater diversity of growth forms (or vegetative types of plants) and a larger number of species than in any other part of arid California. This is the region in which the landscape is dominated by the Joshua tree and in which the ground is frequently well covered by a stand of shrubs and suffrutescent (semiwoody) perennials such as sagebrush, creosote bush, saltbush, bursage, lycium, dalea, hopsage, and others. This region produces a magnificent display of color in the spring of the year. These plants—largely species of *Gilia*—make maximum vegetative growth during the brief period of favorable conditions and depend on an abundant seed crop to continue their existence from year to year.

Toward the Mojave River there is a rapid impoverishment of the vegetation and the flora. At a distance of 30 to 50 miles (48 to 80 km) from the base of the mountains toward the Colorado River, there is a truly desert portion of California in which the vegetation and flora are reduced to extreme simplicity, where creosote bush and bursage form over 95 percent of the plant population. The individuals are often widely separated and of very low stature. The accompanying plants are all of the nonsucculent type, invariably having a very reduced transpiring surface. The ephemeral species are few and likewise of a very xerophytic stamp.

North of Inyo County most of the desert lies above 4,500 feet (1,350 m), and creosote bush and bursage are absent. The most common plant is sagebrush, giving the region identity with the Great Basin type of desert, with which it is continuous to the east. The southern arm includes the Salton Basin and the desert lying between it and the Colorado River.

The northern part of the Mojave Desert resembles the Great Basin Desert in its poor display of life forms, in the simple composition of most of its communities, and in the strong control in distribution of its vegetation by the texture or salt content of the soil. The floras of the two regions have a high percentage of plants in common, but outside the adjacent margin there is a low percentage of species in common. For the most part, the life forms are those of the Great Basin.

Throughout the central and eastern part of the Mojave Desert the surface is covered with a stand of creosote bush and bursage that blankets the desert without reference to drainage, slope, or soil. The only break in the monotony of the plant covering is a slight difference in stature, density, and the irregularity of the spacing that commonly exists on areas of shifting sand. In many large areas creosote bush is reduced to a height of 15 to 18 inches (38 to 45 cm), and the vegetation covers only 3 percent of the surface. The vegetation aside from the dominants is made up of such plants as big galleta, four-winged saltbush, and several minor species of saltbush, joint fir, and some larger shrubs.

The dry lakes usually have a surface that is smooth, hard, and devoid of vegetation. In many cases the prevalent creosote bush and bursage extend to the very edge of the bare lake bed. In other cases there are belts of salt plants.

In the extreme southern tip of Nevada, the low elevations of California, and in part of northwest Arizona the vegetation is similar to that of the Mojave desert farther west but has some characteristics of the Great Basin, as well as a small representation of the plants of the Sonoran Desert.

Cacti in the Mojave Desert are represented by only a few species and by very widely scattered individuals—beavertail prickly pear and silver cholla are the only common ones. For these plants the seasonal distribution of rainfall is particularly disadvantageous. When the soil is moist it is likewise cold and not favorable for the absorption of a sufficient quantity of water to tide over a rainless summer. In southern Arizona, at the end of the arid foresummer, the cacti are frequently in such a state of depletion of their water supply that the abeyance of the customary summer rains would surely result in a serious decimation in their numbers, precisely what would happen to them in the Mojave Desert.

Shreve noted that, with the exception of the immediate vicinity of the Salton Sea, the vegetation of the Mojave Desert is in what is customarily designated as a "very static condition," which means that such physiographic changes as are taking place are not accompanied by changes in the vegetation. This is due to the extreme simplicity of the flora and to the fact that it is composed solely of plants that are all capable of survival in the highly adverse conditions that are presented by all of the local habitats. In a consideration of the dynamic aspects of the vegetation of a region in which the initial, sequential, and final stages of a succession are characterized by the same species (and often by the same individuals), it is doubtful whether the successional conceptions, formed in regions with a very dissimilar vegetation, are of much real utility.

## The Chihuahuan Desert

Shreve and associates made several trips to the Chihuahuan Desert, the last one in 1939. Although the monograph he had planned to prepare on the vegetation of this area was never completed, Shreve's observations were summarized in three articles published in 1942 (Shreve 1942a, 1942b, and 1942c).

The Chihuahuan Desert includes parts of New Mexico, western Texas adjacent to the Río Grande, the lower valley of the Pecos, the eastern half of Chihuahua, the western half of Coahuila, and parts of Durango, Zacatecas, Nuevo León, and San Luis Potosí. A small percentage of the area along the Río Grande is below 3,000 feet (900 m) in elevation, nearly half of the Chihuahuan Desert is over 4,000 feet (1,200 m), and some of the highest parts are over 6,000 feet (1,800 m). Desert plants may often be found at 7,000 to 8,500 feet (2,100 to 2,550 m).

Vast plains and immense undrained basins alternate with mountain ranges rising from 2,000 to 5,000 feet (600 to 1,500 m) above the plains, or with intricate groups of limestone hills. Hard surfaces are prevalent and the only extensive sandy area is immediately south and southwest of Ciudad Juárez in Chihuahua. Limestone is more abundant than in any of the other deserts. Outcrops of granite are uncommon. Deposits of gypsum and soils heavily impregnated with gypsum are widespread.

The rainfall ranges from annual averages of three inches (7.5 cm) in the vicinity of the undrained basins of Coahuila to as much as 12 to 16 inches (30 to 40 cm) at more elevated areas near the western and southern edges of the desert. Nearly everywhere the summer precipitation of June to September is from 65 to 80 percent of the annual total. The precipitation from October to the end of the year is light. The period from January to May is very dry throughout the area. Moderate frosts are common at desert levels or may be severe above 5,500 feet (1,650 m). The summer daytime temperatures are from 10° to 20°F (5.5 to 11.0°C) lower than those of the Sonoran Desert, except at low elevations along the Río Grande. As a result of the summer distribution of rainfall and favorable temperature conditions, the growing season is both longer and more certain than the summer season of plant activity in any other Northern American Desert. There are no winter herbs, no winter germination—and no winter flowering—although some plants remain wholly or partly in leaf during the cold months.

The number of life forms in the Chihuahuan Desert is greater than that of the Great Basin and less than that of the Sonoran Desert. There are relatively few cacti as compared with the Sonoran Desert but more grasses. Shrubs and semishrubs are the predominant plants. Trees are small, few, and confined to the margins of streamways or to rocky slopes. Stem succulents from 6 to 12 inches (15 to 30 cm) are very common, two species of chollas are widespread and locally very abundant, prickly pears are locally common, and two large species of barrel cacti are more conspicuous than abundant. The columnar type of cactus barely enters the southern end of this desert. The semisucculents, palmilla, beargrass, and sotol are the most conspicuous of the larger plants and are accurate indicators of habitat conditions. Agave and mescalito occur widely and are particularly abundant. The ocotillo is found throughout the Chihuahuan Desert. In certain localities, grasses are abundant in the upper margin of the desert, but in other places they fail to appear until an elevation is reached above the vertical limit of the desert shrubs and cacti.

The Sonoran and Chihuahuan deserts have only three species in common among the most characteristic plants: creosote bush, mesquite,* and ocotillo. The legume family and the sunflower family are both richly represented in both deserts, but the former family appears to be better represented in the Sonoran Desert and the latter in the Chihuahuan.

The highlands of western Chihuahua are the source of three rivers that flow north into the extreme northern part of the state and in large lake beds. Surrounding these beds is a sandy area about 100 miles square (259 square km). Very little of this is stabilized and much of it is occupied by active dunes from 50 to 300 feet (15 to 90 m) in height.

From the dune region east to the valley of the Río Conchos the vegetation of the Chihuahuan Desert is a uniform and monotonous stand of creosote bush, or sometimes creosote bush, tarbush, and Chihuahua whitethorn with broad low buried trees of mesquite on the low ground. Spiny allthorn is abundant, cacti are frequent but not conspicuously abundant, and yuccas are not common.

East of the Río Conchos an increasing percentage of the surface is occupied by hills and small mountain ranges, in which granite, sandstone, and limestone are represented, and the general elevation of the intermontane plains reaches 5,000 feet (1,500 m). In this area the desert merges irregularly into grassland or into evergreen oak shrubbery, the boundary of the desert being in northern Coahuila.

The grasses are an infrequent element in the driest part of the Sonoran and Mojave deserts. When they cover more than 50 percent of the surface, shrubs and cacti are very widely spaced, and, when some of the desert dominants have disap-

*Some botanists believe the mesquite in the Chihuahuan Desert is a distinct species, different from that in the Sonoran Desert.

peared, it is an indication that the transition to grassland has been entered. In northern Chihuahua there are a number of extensive level plains that are heavily and exclusively covered with grasses, almost solely tobosa grass. These *llanos* occupy the deep fine soil of the bottoms of undrained basins into which there has not been sufficient drainage to result in the development of a central alkaline dry lake. Although grasses are the dominant species, these areas should be considered as a desert association.

The southern arm of the desert is much more arid than the southwestern. The great open expanse of northern San Luis Potosí has an annual rainfall of about four inches (10 cm). There is no outlet for the drainage and the soil is everywhere very alkaline or else has a very high content of gypsum. There are large tracts covered with creosote bush from 2 to 3 feet (60 to 90 cm) high, either nearly pure or competing with tarbush in places where the soil moisture is sometimes higher. Very open forests of mesquite occupy favorable sites, with a heavy ground cover of saltbush.

*It is true here, and it is throughout the desert, that the most widely distributed and drought-resistant plants are the ones that dominate the innermost plains. In every part of the desert it is the plants of the upper bajadas and hills that give the vegetation of the region its distinctive character and include in their number species of a special interest on account of their structural modifications, relationships, or geographical range. Also, on approaching the edge of the desert or one of its subdivisions, it is the upland rather than the valley plants that give the first suggestion of impending change. [Shreve, 1942a]*

## Deserts of Southern Mexico

South of the Chihuahuan Desert there are discontinuous areas in which vegetation differs widely from that of the three subdivisions already described. Nevertheless, these areas exhibit the essential characteristics of less extreme types of desert. They are rich and diversified, with a large number of life forms and many plants of distinctive habit, structure, and behavior. The stature and density are very irregular but exceed those of the northern sections of the desert.

The southern desert areas are closely related to the arid bushland and to the thorn forest, and they merge gradually into these formations. Desert plants and ecological features of the desert recur again and again in the rugged terrain of southern Mexico, although often in areas that cannot be regarded as desert.

In northern Mexico the broad valley floors have the most simple and drought-resistant vegetation. In southern Mexico the valley floors are again the most arid habitat and the one in which vegetation most nearly resembles that in the north. Throughout Hidalgo, Puebla, and adjacent states, a very slight amelioration of the soil and soil moisture conditions suffices to support arid bushland, thorn forest, or other types of xeric or semixeric vegetation. In and around the arid valleys are hills, canyons, barrancas, and narrow floodplains on which a more mesic vegetation is dominant.

All of the life forms found in the Sonoran and Chihuahuan deserts are represented in the southern desert areas. The features of form, habit, structure, and behavior that are the distinctive characteristics of desert plants are found in the southern areas, either in their initial or final phase of development. Also, many of the genera that have contributed one or two highly specialized forms to the pronouncedly desert areas are represented in the south by a large number of species. In the members of these genera may be detected structural features that have undergone further development in certain species, making life possible for them in the very arid regions.

The small size of leaves or leaflets is a very common feature of the plants of arid southern Mexico. The assumption of photosynthetic work by the stem is common, although not often accompanied by total loss of the leaf. Plants with succulent leaves are even more abundant in the south than in the north. Semisucculent plants are abundant in species in the south but not so dominant in the vegetation. The stem succulent cacti are perhaps even more abundantly represented in both species and individuals in the south than they are in the Sonoran and Chihuahuan deserts.

In considering the origin of life forms of flora in the northern and southern parts of the North American deserts, it is important to note that the northern part of the desert is nearly surrounded by heavily forested mountains in which all of the habitats are favorable to mesic plants. To the south, however, desert merges gradually into thorn forest, arid grassland, or semiarid shrubland. There are no barriers to interchange between these plant formations.

With this understanding of some of the general aspects of deserts, it is possible to address the subject which preempted the attention of the Desert Laboratory pioneers for thirty-eight years: the Sonoran Desert.

# The Sonoran Desert

## *General Description*

THE MOST COMPREHENSIVE STUDIES of the Sonoran Desert were made by Shreve and his associates in the years following 1932; they resulted in the publication *Vegetation and Flora of the Sonoran Desert*—the first part, *Vegetation of the Sonoran Desert*, was published in 1951, a year after Shreve's death.

Preparation of the manuscript had been delayed by the closing of the Desert Laboratory and by World War II, and later by Shreve's failing health. The final two volumes, including Shreve's original text and a complete description and catalog of the flora, added by Ira Wiggins, were published in 1964 (Shreve and Wiggins, 1964).

Shreve's descriptions are comprehensive and detailed. He covers climate, physiography, and soils. His observations on the flora are also extensive, and he includes lists of important plants.

Valuable as Shreve's material is for the student or specialist, it has been the intention to present it here in a form appropriate for the general reader who wants accurate but easily absorbed information about the Sonoran Desert as one of the most diverse and interesting areas on the globe. The descriptions of vegetation that follow are largely based on material in the *Vegetation and Flora of the Sonoran Desert* but also include highlights from many publications by Shreve and other Desert Laboratory scientists.

The boundaries of the Sonoran Desert are sharply defined wherever topography change is abrupt (i.e., in Arizona between Safford and Wickenberg, in California along the east base of the San Jacinto and Cuyamaca mountains, in Baja California along the east foot of the Sierra Juarez and Sierra San Pedro Martir, and in Sonora near the confluence of the Bavispe and Moctezume rivers).

In level or rolling regions the boundary is ill defined since there is a gradual transition from desert to adjacent types of vegetation (i.e., in Sonora between Quiriego and the Gulf of California, in Arizona and California in the vicinity of Needles, and in Baja California to the north of Rosario). From the headwaters of the Salt River in Arizona, southward to Moctezume, Sonora, the boundary lies at an elevation of 3,000 to 3,150 feet (900 to 945 m). All other sections of the boundary lie at lower levels. The well-known influence of slope exposure in determining the distribution of vegetation is responsible for many tongues and islands of desert that give irregularity to the boundary in its rugged sections.

The approximate area of the Sonoran Desert is 119,370 square miles (310,362 square km). This is divided among the four states as follows: Sonora, 48,560 square miles (126,256 square km); Arizona, 40,450 square miles (150,404 square km); Baja California, 24,104 square miles (62,670 square km); California, 6,166 square miles (16,032 square km).

Nearly all of the Sonoran Desert drains into the Gulf of California; a very small fraction drains into the Pacific Ocean. The largest rivers rise outside the deserts in the mountains of Colorado, Arizona, and Sonora. There are a number of small undrained basins at low elevations in Arizona, Sonora, and Baja California, including the Salton Sea in California and the Laguna Chapala in Baja California. The only river with a perennial surface flow is the Colorado. Before dams were built on the Salt and Gila rivers, the Gila was a nearly constant stream. The largest endogenous desert river is the Sonoita in northwestern Sonora. Its floods and those of the Río Sonora are never discharged in tide water. The Río Magdalena rarely discharges its flood waters to the Gulf, but originally the Río Yaqui did so annually. In the rainy season these desert tributaries are in flood a few hours or sometimes a few days. The Sonoran Desert is hedged in by mountains that rise gently or abruptly from its borders except in three instances: (1) along the line of separation from the Mojave Desert between Needles and Indio, California; (2) along the line of separation from the cape region on the Pacific side of the tip of Baja California, and (3) along the line between desert and thorn forest in southern Sonora. Although there are many mountains and hills in the desert, only a few are high enough to bear mesic vegetation.

The mountains wholly within the Sonoran Desert are generally of such low elevation that their summits are clothed solely by desert plants. The mountains in the California section of the Sonoran Desert and in the extreme southwestern part of Arizona appear to be devoid of vegetation, but they do support an extremely open stand of highly xeric plants not abundant enough to influence the color given the slopes by the bare rock. In Arizona and Baja California there are numerous volcanic mountains with extruded plugs or basaltic mesas contributing to their bizarre outlines. The abrupt walls, bare columns, and deep canyons found in small mountains of this type are well developed in the Ajo, Castle Dome, and Kofa mountains in Arizona; in the Sierra Viejo in Sonora; and the Sierra San Borja and the Tres Vírgenes and Santa Clara mountains in Baja California.

The climate of the Sonoran Desert is relatively uniform with regional differences due to latitude, elevation, and the geographical configuration. The climate is distinctly of the continental type, especially in the northernmost part. In the southern portion, the influences of the sea include a slight lowering of the diurnal temperature, the intermittent effect of fog, and the almost constant blowing of a strong, onshore wind. In Baja California the main effect of the ocean is a slight lowering of the sensible temperature within 25 miles (40 km) off the shore during the daytime. All of the slopes in Baja California are relatively cool owing to onshore breezes.

The winter temperatures vary considerably from north to

south, but summer temperatures change little because the influence of latitude is outweighed by the more continental position of the northern end. The gradual increase in elevation from the coasts to the interior results in a slight fall in the temperature range with occasional severe frosts at the upper elevations. Also at the upper elevations there is a considerable increase of precipitation in the interior, augmented by the mountain wall that forms the northern and eastern boundary.

Shreve noted that parts of the Sonoran Desert share with Death Valley the highest and most sustained air temperatures in North America. Periods of ninety consecutive days with a maximum of 100°F (38°C) are common. And even in February, in the winter, the temperature can get up to 90°F (32°C).

There is probably some vegetative activity in local favorable spots throughout the year. The length of the frostless season ranges from eight to twelve months: however, the duration of the potential growing season is interrupted by rainless periods that sharply reduce all plant activity.

Shreve (1951) subdivided the Sonoran Desert into seven areas: the Lower Colorado Valley, the Arizona Upland, the Plains of Sonora, the Foothills of Sonora, the Central Gulf Coast of Sonora, the Vizcaíno Region, and the Magdalena Region. Areas adjacent to and overlapping Shreve's seven original categories offer much information about the origins and development of desert species. For this reason, the Cape Region is added here, as well as Baja California as described in portions of Shreve's diary.

The descriptions of vegetation that follow are largely based on material in the *Vegetation and Flora of the Sonoran Desert* but also include highlights from many publications by Shreve and other Desert Laboratory scientists.

## Lower Colorado Valley

The Lower Colorado Valley, extending westward from Phoenix and Ajo in Arizona to Indio in California, is the largest subdivision of the Sonoran Desert—400 miles (640 km) in length. It includes the lower drainage of the Colorado and Gila rivers, the Salton Basin, the east coast of Baja California as far

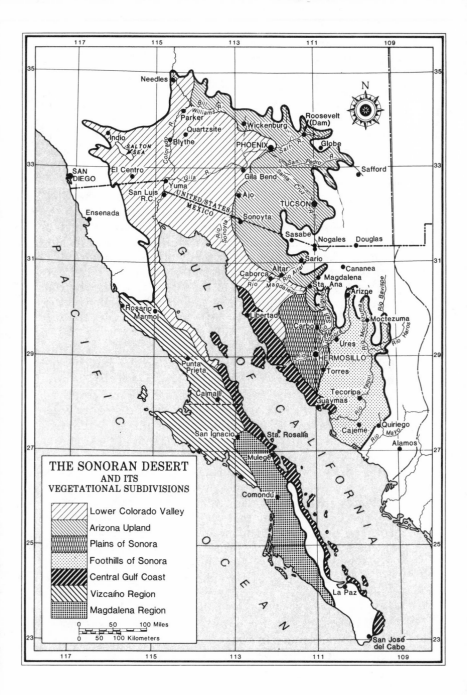

THE SONORAN DESERT
AND ITS
VEGETATIONAL SUBDIVISIONS

Lower Colorado Valley
Arizona Upland
Plains of Sonora
Foothills of Sonora
Central Gulf Coast
Vizcaíno Region
Magdalena Region

0        50      100 Miles
0      50     100 Kilometers

south as Bahía Los Angeles, and all of Sonora lying below 1,300 feet (390 m) elevation as far south as the valley of the Río Magdalena. It is chiefly drained by the Colorado River, but its two southern extensions contribute their floodwaters directly to the Gulf. The Lower Colorado Valley is well defined along its western edge and rather sharply marked between Parker and Phoenix as well as between Sonoita and Santa Ana. Elsewhere, the vegetation spreads gradually into adjacent subdivisions.

An area of fewer hills and mountains than the other divisions, the Lower Colorado is one of the most arid parts of the Sonoran Desert—low in rainfall and soil moisture and high in sunshine, both summer and winter. Surface features include volcanic and granitic mountains, sandy plains and dunes, alkaline flats and soils varying in origin, texture, and depth. However, it does include the delta of the Colorado River where an abundant moisture supply once gave rise to a series of vast aquatic communities that had little in common with the surrounding desert vegetation (Sykes, 1937a).

The vegetation of the Lower Colorado Valley is distinguished by its simplicity of composition on the gravelly and sandy plains which make up a high percentage of the the area. The upper bajadas, the level fields of recent volcanic eruptives, or malpais, and the hills and mountains support a richer vegetation. The two species that outnumber all others are the creosote bush and white bursage. Low, open stands of these shrubs make up 90–95 percent of the plant population over large areas. On the rocky soil of upper bajadas and pediments are to be found small trees, notably mesquite and foothill paloverde.

Plants are more abundant along the larger drainageways. Also, species occur there that are absent from the level surfaces. The trees most abundant are mesquite, blue paloverde, ironwood, and smoke tree. The size of the trees is roughly proportional to the size of the drainageway. Along the smallest washes they may occur as mere shrubs or may be absent. The most common plants along the small drainageways are burro brush, seep willow, Anderson lycium, and catclaw.

Compensation for the poverty of perennial plants among the vegetation is found in a large flora of ephemeral herbaceous species, especially following the late winter rains and particularly on sandy soil. Winter annuals are much more abundant than in any of the other regions of the Sonoran Desert but less abundant thañ in adjacent portions of California. Creosote bush and white bursage retain their dominance on soils that differ greatly in depth and texture. This is in contrast with their restricted distribution where rainfall is greater. These two plants are either dominant or abundant under differences of substratum which would support dissimilar vegetation in more favorable moisture conditions. The simplicity of flora means merely that no other shrubs have developed biological and physiological characteristics as highly suited to survival in this region as are those of creosote bush and white bursage.

### Volcanic Mountains and Basaltic Mesas

The Lower Colorado Valley includes many areas of varying size and topography occupied by volcanic eruptives which either are recent or have undergone very little erosion or degradation (Shreve, 1951; Shreve and Wiggins, 1964). The largest mountain in this group is Pinacate Peak in Sonora, Mexico. It is surrounded by smaller peaks, volcanic plugs, and an irregular lava field 31 by 46 miles (50 by 74 km) in maximum extent. This field is the "lava" of Hornaday's *Campfires on Desert and Lava* (Hornaday, 1908). The largest recent volcanic rock, or malpais, area in the United States is located near Sentinel, Arizona.

The volcanic surfaces are exposed in every position from vertical cliffs to nearly level areas. A very common form is the gently tilted mesa, extending for several miles at the same gradient and ending as a scarp at the high end and along one or both of the sides. In southern Yuma County and western Pima County, Arizona, many of the basaltic hills consist of older material and are merely covered with a veneer of basalt from 3 to 6 feet (1 to 2 m) in thickness.

Throughout the extreme northwestern corner of Sonora, sand and basaltic rock are often found close together. Accumulations of sand are common at the windward base of hills and have encroached on some of the nearly level malpais fields. The volcanic rocks have a veneer which is very dark brown or almost black and polished by the moving sand. The result is a strong color contrast in the landscape, as well as a marked dissimilarity of vegetation.

The Pinacate Mountains were studied by MacDougal and others in the expedition graphically described by Hornaday. This area includes over two hundred volcanic cones with an abundant vegetation and some plants of the Tucson region, including the saguaro, creosote bush, brittle bush, cholla, and a few others, descending to nearly sea level.

According to Shreve the vegetation on the volcanic areas is determined by three sets of soil conditions associated with the types of surface. On the edges of flat areas of explosive origin, the conditions of the plain are slightly modified in the direction of conservation of moisture. On the thin but nearly continuous veneer of impervious rock covering a retentive soil, runoff penetrates the cracks in the lava and is retained by the soil with extremely small loss by evaporation. On the thick beds of basalt, a very fine brown clay is formed by weathering and accumulates in pockets of various sizes. Vegetation indicates the slight improvement in moisture retention around the edges of the malpais fields and very much better conditions existing on the thick beds of lava.

The extremely rocky and rough surface of the volcanic hills, tilted mesas, and level malpais areas exhibit a great lack of uniformity in their plant cover. There are about 20 trees to 2.5 acres (1 ha) in the same localities and none in others. Shrubs and semishrubs may grow abundantly or may be so few that the bare areas of black-brown rocks give their color to the landscape. Trees are more abundant on small hills and level malpais than on large hills, and semishrubs are commonly more abundant where the rocks are largest and the surface most broken. Vegetation is usually denser at the base of a hill than on its upper slopes, and invariably taller in the shallow canyons and waterways. On the upper slopes there are frequently extensive areas that are nearly bare, or stretches of loose rock, with a depth of three or more feet (1 or more m), with no soil and no vegetation other than the crustose lichens covering rocks.

Shreve (1951) made a distinction between "older" and "recent" on the basis of the stage of physiographic development of their surfaces rather than their historical sequence.

Some of the "older volcanics" may be contemporaneous with the unmodified areas of basalt. The mountains and hills of the older volcanics are those built of rhyolite, andesite, basaltic conglomerate, and all other volcanics except recent basalt. Mountains of this class are highly eroded, usually with bare peaks or summits and with very narrow lateral ridges. The surface is either rock in place or made up of an accumulation of coarse and unstable angular stones, with a poor development of soil. Since all the mountains and hills of this type in the Lower Colorado lie in the region with less than 6 inches (150 mm) of rainfall, the conditions for vegetation are extremely unfavorable.

Viewed from a distance of 1 or 2 miles (2 or 3 km), the older volcanic ranges appear to be entirely devoid of plants. On closer approach it is found that there is a very scattered and irregular cover of small perennials and grasses. Even in the canyon bottoms of the ranges near the Colorado River there are few trees, except where a local accumulation of sand or clay has favored the blue paloverde or ironwood.

In many favorable places the brittle bush is conspicuous on the slopes of the mountains, and the ocotillo occurs consistently on the outwash slopes below. In extreme southern Arizona, organ-pipe cactus appears on rocky slopes, usually on a substratum such that it is accompanied by ironwood and ocotillo. One of the most ubiquitous plants on the bare slopes is tanglehead grass. The only cacti that are frequent are California barrel cactus and Johnson echinomastus.

### Delta of the Colorado River

MacDougal and Sykes were greatly interested in the delta of the Colorado River, which has much in common with lower deltas in other arid regions of the world. Their studies were opportune as the building of dams in the late thirties changed the entire regime of the Colorado River. Processes that had been going on for centuries—flooding and silting that had been a feature of delta formation and maintenance—were brought to a stop. By 1980 the whole aspect of the area had been changed by the effect of upstream dams and the ever encroaching settlements from farming and recreational uses. MacDougal (1904b) pointed out the difficulties of travel and subsistence near the mouth of the Colorado River in the early part of the twentieth century. The southern part of the delta included vast areas of muddy salt flats cut by a labyrinth of shallow pools and channels, and joining directly the desert slopes and plains of Baja California and Sonora. The water in the lower course of the river was brackish for a distance of 20 miles (32 km) from the sea, and other sources of water were uncertain and widely separated. In running the boundary on the long northwestward slant of the Arizona-Sonora line after the Gadsen Purchase Treaty, the commission found it necessary to haul water nearly 124 miles (198 km) to meet the needs of its camps.

MacDougal organized an expedition to this region early in 1904, under the joint auspices of the Desert Botanical Laboratory of the Carnegie Institution and the New York Botanical Garden. In the preceding November Sykes had gone to Yuma, where he had constructed a small sloop, 30 feet (9 m) in length with an eight-foot (2.4 m) beam. This boat was flat-bottomed in design—suitable for floating down the the muddy shallows of the river—was rigged with a mainsail and jib and furnished with a center board for use in sailing the rougher waters of the Gulf.

Near the course of the river in 1904, the delta consisted of an alluvial plain—not more than 13 feet (4 m) above the low-water mark—subject to constant bank erosion, shifting, and remaking of the soil. The area was cut in all directions by old channels existing as bayous and sloughs and flooded at high water in May, June, and July. Almost pure formations of willow and cottonwood covered many square miles and furnished food for thousands of beavers that burrowed in the banks.

Large areas were occupied by arrowweed, mesquite, and screw bean. Two or three species of saltbush were found where the action of the water prevented the establishment of the woody perennials of greater size. In the upper part of the delta the common reed fringed the channel, and its closely interwoven roots acted materially in preventing erosion of the banks. In the lower part of the delta, where the river was affected by the spring tides, the common reed was partly replaced by cattail, which not only lined the shores for many miles, but extended back some distance on areas free from trees, forming dense masses that afforded shelter for a number of animals including a peculiar small mountain lion subspecies. The forest of willow and cottonwood began to lose density at a distance of 30 to 35 miles (48 to 56 km) from the Gulf, the willows extending farthest toward salt water. Beyond these were the mud plains, portions not actually subject to erosion because of a thick covering of salt grass and small clumps and isolated specimens of saltbush, mesquite, and screw bean. Such areas became inundated at the highest tides; consequently the soil solutions were heavily charged with salts, and whitish alkaline crusts appeared during the winter dry season.

The floods of spring and early summer from the rains and melting snows of the headwaters region of the river raised the level of the water until it flushed the innumerable old channels and covered the greater part of the delta. Most of the herbaceous species made their annual growth after the waters subsided in July. Other species, less affected by the lower temperatures and low relative humidity of the winter season, were set in action by the favorable conditions of March and April and came into bloom at this time, thus making two distinct seasonal groups of annuals.

The gravel plains or mesas showed striking contrasts with the pure dense formations of the muddy soil of the alluvial plain of the delta. In places the creosote bush descended the gentler slopes to the margin of the moister soil near the margin of the channel, with a height of over 23 feet (7 m), the maximum size for the species.

The elevations included in the delta were dry mountain slopes that support a desert vegetation. The mesas adjoining the northern part of the Gulf of California appeared to offer the most extreme desert condition in North America. The rainfall at Yuma in the northern extremity of the delta was less than 1 inch (25 mm) during 1903 and, at points farther south, years have been noted in which no precipitation occurred.

MacDougal added further vivid descriptive material on this area:

*The Rio Colorado, the Nile of America, rises in the higher mountains of Wyoming and Colorado, cuts its way through canyons, mile deep for hundreds of miles, to the western edge of the great plateau, where it emerges to*

*flow through the most arid desert in North America, finding its way into
the head of the Gulf of California in a sub-tropical climate two thousand miles
[3200 km] from its source. Here it spreads its delta with one edge to the
westward against the huge sand-dunes or "algodones" of southern California, while the other side of the fan lies far to the eastward along the margin
of the Sonora mesa, where even the jungle of the swamp meets the sparse
vegetation of the arid gravel slopes and bluffs in striking contrast. Between
the deserts to the west and to the east the radiating ribs of the delta fan,
variously intertwined with subdivisions of the river, project against the
Cucopa Mountains, which now rise clear-cut and cameo-like and now loom
hazily against the southern horizon. The delta and the range beyond it, like
many such "unknown" regions, have been visited in desultory fashion by
the prospector, the hunter and trapper, the nearest camps teem with stories
of adventures, hazards, and disasters among its sloughs, tidal bores, vol-
canoes, sunken saline plains, and arid mountains.*

Before the construction of Hoover Dam, an immense
amount of sediment, estimated at 60 million tons (54 million
MT) yearly, was carried down into the delta and deposited on
the bars and banks of the various channels, giving rise to the
assertion that appreciable construction of solid land has occurred within the observation of persons now living. This view
is supported by the fact that the willows and poplars, which
do not inhabit the lowest land moistened with brackish water,
have advanced about 10 miles (16 km) farther south within the
memory of certain Indians, and also by the formation of Hilda
Island in the lower reach of the river, first noted by Sykes while
descending the river in the sloop *Hilda* in 1890. Sykes (1937a)
gave an excellent historical summary of the delta, beginning as
early as 1539 with the exploration of Francisco de Ulloa. The
description includes copies of some of the earlier maps of great
historical interest. He gives credit to Father Kino for finally
exploding the misconception concerning the Gulf of California. Although Kino's descriptions were not too complete, they
very definitely showed that Baja California was a peninsula,
rather than an island.

Sykes (1937b) also documented the changes resulting
from the completion of Hoover Dam which inaugurated a new
regime for the downstream section of the Colorado. The violent fluctuations in water volume, both annual and seasonal,

caused by the varied climatic environments from which the waters of the source streams and other tributaries were derived, no longer occurred in the lower river. The delta, therefore, entered a period of relative fixity, and channel changes became slight. The vegetation entered upon a cycle of development and an associational history that of necessity was markedly different from what it would have experienced had the river been allowed to flow naturally.

Striking as have been the past constructional activities of the river in relation to the sinking trough, the latest chapter of the river's history, discussed by Sykes, is no less interesting. After delivering annually for thousands of years tremendous quantities of sediments derived from the Rocky Mountains and the vast intervening plateau areas, the Colorado River was suddenly brought under control, its floods were checked, and the supply of sediments from all but the lower part of its course was abruptly cut off. The changes in the stream's habits with reference to the sediments remaining available in its lower channel—its scouring, filling, sorting, and depositing—form a subject of great scientific interest and significance in dynamical geology. These changes can have very important relations to the utilization of the lower river for irrigation and municipal water supply, in which huge sums have been invested.

The general changes toward a condition of desiccation in the delta were soon quite noticeable—in the indigenous fauna of the region as well as in the vegetation.

The dearth of animal life was very striking. Freshwater fish, formerly existing in such incredible quantities in the channel and in every other body of river water above the tidal influence, and the saltwater fish that found such abundant food supplies about the river's mouth, were so greatly decreased in numbers as to be almost nonexistent by comparison.

Bird life was greatly diminished; the waterfowl, so constantly in evidence along every channel and backwater, became far less numerous as their food supply disappeared. And many mammals, such as the raccoon, beaver, and water rats that formerly were so plentiful in all the perennially running water, also practically disappeared.

## Arizona Upland

The vegetation of the Arizona Upland makes this area perhaps the most interesting of all desert areas in the world. There are more large perennials per unit area than in any other group of habitats, the stature of the large perennials is greater, the number of associated, smaller perennials is greater, and the species vary more. The most abundant species, present in nearly every square mile, include creosote bush, foothill paloverde, staghorn cholla, Engelmann prickly pear, ocotillo, Anderson lycium, lotebush, desert hackberry, and velvet mesquite. Saguaro is far less numerous, but height makes it a conspicuous element in the vegetation. Ocotillo is particularly abundant on the pediments and very coarse soil. Creosote bush is more abundant on level areas and benches than on the steeper slopes with coarse, shallow soil. As the bajada is ascended, foothill paloverde increases in abundance, and mesquite more frequently reaches the size of a small tree. The southern end of the Arizona Upland in Sonora, Mexico, is similar in physiography and vegetation, but the density is slightly greater. The flora, while dominated by the same species, includes some that are not found in the northern part.

Coville and MacDougal were so charmed by the Arizona Upland of the Sonoran Desert that they chose it for the site of the Desert Laboratory. The Sonoran Desert has been well loved and vividly pictured by authors like Joseph Wood Krutch, *The Desert Year* and *The Voice of the Desert,* and Mary Austin, *The Land of Journey's Ending.* Glowing words are to be expected from these writers, but scientists—in ecology, meteorology, zoology, and plant physiology—loved the area and wrote fascinating descriptions as well.

*A brisk ride of about four miles [6 km] downgrade from the summit of Roble's Pass (just west of Tucson) brings the outfit down to the main floor of the desert, at the eastern edge of the Avra Valley. But where is the barren, lifeless waste of drifting sand, desolation and danger that naturally rises in the mind of the uninitiated reader whenever deserts are mentioned? Clearly, it is not here.*

*We see ahead of us, stretching away mile after mile to far distant ranges of hazy-blue mountains, a vast plain, level as a race-course, but completely covered with cheerful-looking verdure growing about waist high to a man on foot. Instead of being a gray and melancholy waste, however, like the sagebrush flats of Montana and Wyoming, this great garden is green —persistently, cheerfully, even delightfully green! And you do not see anywhere even so much as half an acre [0.2 ha] of perfectly bare and verdureless ground. There is bare ground between these green clumps of creosote and mesquite bushes; but that is only a bit of novelty in Nature's planting scheme.*

*How very unlike the desert of our expectations. Let us call it, for truth's sake, an arboreal desert. . . .*

*Strange to say, there is in those gray mountain walls a sense of cheerful companionship that quite robs the deserts of the awful monotony that usually characterizes uninhabited level plains of illimitable extent. To some minds the idea may seem absurd, but to me the mountain ranges were company. The ranges near at hand are always so isolated, so sharply defined, and so individualized that they are as much company to the wayfarer as so many houses with windows that look at you. To perish on a great waste of sand like the Sahara would be very monotonous and disagreeable, but in one of these beautiful green plains, surrounded by an amphitheatre of interesting mountains, death would be quite a different matter.* [Hornaday, Campfires on Desert and Lava]

Ellsworth Huntington (1911), noted climatologist and geographer, was greatly impressed after a brief visit to the Desert Laboratory and the Sonoran Desert, and he expressed it as follows in *Harper's Magazine:*

*It is not the desert of our childish fancy, with drifting yellow dunes, and five straight palm-trees standing beside a pool, to which an Arab horseman, spear in hand, stoops for a drink. Nor is it the barren gravelly desert of Persia, most desolate of all the landscapes of the world; nor yet the drab sagebrush desert of Utah and Nevada. On the contrary, it is the most beautiful of deserts, the strange arboreal desert of southern Arizona. . . .*

*It would be hard to find any flowers more beautiful than those which form a coronet of white around the tops of the giant cactus, and produce a many-seeded fig-like fruit, one of the staples of the diet of the Indians. The slender bells some three inches [7.6 cm] long, are not pure white, but of a slightly creamy tint. Heavy masses of yellow stamens form a pleasing contrast to the pale petals, especially when a shiny black bee is burrowing for honey. The petals have a peculiarly soft quality, not sticky, but as if the surface had actual depth. The low growing pricklypear is by no means as aristocratic as the saguaro, but its wide-open yellow blossoms, shading sometimes to lemon and sometimes to orange, have a very friendly quality. In the flowering season no cacti are so interesting as two closely similar species . . . known to scientists as staghorn cholla and cane cholla. They are scraggly, branching forms from three to six feet high, [1 to 2 m] with stems of many shades from purple to green. Ordinarily they are unattractive, but when the flowers come out, one is tempted to spend hours in wandering from one to another to see what the color will be. Each plant has its own peculiar color, the same when newly opened or when fading. On some plants the flowers are almost green, on others pale yellow, orange, or brown. And as if this were not enough, one soon finds plants whose blossoms are bright pink or purple, or varying shades of red. A single plant never has flowers of more than one color, but one may look at scores of different plants and scarcely find two bearing the same shade.*

The Arizona Upland includes the northeastern part of the Sonoran Desert, mainly Arizona, partly in northern Sonora, Mexico, and lying at elevations between 500 feet (150 m) and 3,000 feet (900 m). On the south and west the area merges into the plains of the Lower Colorado Valley; on the north and east it is flanked by a broken series of mountains, most of which rise above the level of desert conditions and vegetation. Precipitation ranges from nine. inches (22.5 cm) on the west to 13 inches (32.5 cm) on the east at higher elevations. The limit of the vegetation is clearly marked, although a few of its characteristic plants occur as high as 6,000 feet (1,800 m) in the mountains. This is essentially the portion of the Sonoran Desert east of Wickenburg, Phoenix, and Ajo in Arizona and extending to the eastern limits near Safford and Tucson. It includes an extension into Sonora as far south as Carbo.

This subdivision has a minimum north-and-south length of 400 miles (640 km) which is nearly as great as that of the Lower Colorado Valley, yet there is more uniformity in the composition of the vegetation in the former. Owing to the physical features, the vegetation in the Arizona Upland is relatively homogeneous. There are few sharp contrasts between the vegetation of volcanic hills and soils and that of granitic ones with this exception: the differentiation of the vegetation can be attributed to gradient of slope, depth of soil, physical texture of soil, and surface conditions affecting infiltration. These conditions are closely correlated with physiography and result in the consistent repetition throughout the area of nearly similar habitats occupied by similar vegetation.

The vegetation of the Arizona Upland exceeds that of the Lower Colorado Valley in stature, density, and the number of species that play a dominant or subdominant role. Creosote bush is an important plant but has a greater number of associates. Foothill paloverde, though absent from the lower bajadas, is so abundant as to characterize the area. Other important species are mesquite, ironwood, and ocotillo. Cacti are numerous, the most conspicuous ones being saguaro and barrel cactus and a wide variety of prickly pears and chollas. Perennial grasses are locally abundant; both winter and summer ephemerals carpet the ground in favorable seasons. As contrasted with the Lower Colorado Valley, there are fewer species of winter annuals and more of summer annuals.

In the broadest valleys the vegetation is almost as open as in the Lower Colorado Valley, but in other areas the cover is more dense, occupying 20 to 60 percent of the surface. Creosote bush is abundant in the broad valleys, with white bursage and triangle leaf bursage, which is unimportant in the Lower Colorado Valley, occupying a subdominant place. On coarse outwash plains and slopes triangle leaf bursage is dominant among the individuals forming the matrix in which creosote bush is likewise abundant.

The paloverde-cactus desert of the Arizona Upland occupies the inner and more elevated part of the northern half of the Sonoran Desert. Its greatest width is 155 miles (248 km) in the latitude of Tucson. In its northwestern extension along the foothills of the Colorado Plateau and in its southern extension into Sonora, the area narrows to a width of 30 miles (48 km) or less. The elevation rises gradually from 500 feet (150 m) at Parker and 1,000 feet (300 m) at Phoenix, Arizona, to 3,000 feet (950 m) at sections of its boundary near Tucson, Arizona, and Magdalena, Sonora. Approximately 25 percent of the total area is occupied by hills and mountains. North of the international boundary the drainage of the Arizona Upland is into tributaries of the Colorado; south of the boundary it is carried by Río Magdalena, Río Sonora, and their tributaries. Along the lower courses of the Santa Cruz and Gila rivers there is a large area of outwash plain in which hills are few and widely spaced.

Through the plains and lower bajadas the uniform appearance of the vegetation is sometimes broken by isolated plant occurrences of species that are abundant on the upper bajadas, especially ocotillo, barrel cactus, saguaro, lotebush, and desert hackberry. Some plants not found in other habitats in the region include the crucifixion thorns.

In a report in *Carnegie Yearbook 36* on a trip to the northern end of the Sonoran Desert, Shreve stated that the northernmost occurrence of typical assemblages of Sonoran Desert plants is in the immediate vicinity of the Colorado River between Searchlight Ferry and Hoover Dam. The combination of species may be of more interest to the ecologist than to the casual observer, but this particular area appears to be the meeting place between plants of the Mojave and Sonoran deserts, along with intrusions from the adjacent woodland. The northern limit of the saguaro was found to be at the southern end of the Hualapai Mountains, where it grows with junipers, as it also does on part of its eastern limit in Sonora. The Joshua tree, always thought to be unique to the Mojave Desert, was followed for 150 miles (240 km) in Arizona, and its southern limit was established in a region with climatic conditions very different from those at the northwestern limit of its occurrence in California.

Between the Gila and Salt rivers south of Chandler, the plains are nearly level with a deep loam soil. This area has been largely under cultivation, since World War I, but it originally bore a heavy cover of desert saltbush and an open forest of saguaro. This locality is the only one in which saguaro is known to have occurred abundantly on a relatively fine, rock-free soil.

Approximately half the area of innermost plain in the Arizona Upland is designated as upper bajada. It includes the pediments that often surround the mountains, and a varying percentage of the outwash slopes that fall at uniform gradient toward the central drainageways. Where the mountains are widely spaced, the zone of upper bajada may cover only one-fifth of the distance from mountain to drainageway, but, where they are closely spaced, the physical and vegetational features may extend down the entire slope to its base. The boundary zone of integration between upper and lower is very broad and indefinite. Also the vegetation of the upper bajada, especially where bedrock is shallow, approaches closely that found on the slopes of hills and mountains.

The areas of upper bajada encircle all the isolated mountains in the central part of the Arizona Upland and include nearly all the innermost areas south and west of the Salt River and north of the Altar River. Contrasted with the mixed material of the soil of the lower bajadas, the upper soil is derived from a single mountain range, or part of such mountain, and commonly has distinctive physical features attributable to the character of the rock from which it is derived.

The lower slopes of the mountains throughout the central part of the Arizona Upland are dominated by foothill paloverde, ocotillo, saguaro, brittlebush, whitethorn, and several chollas and prickly pears. On upper slopes blue paloverde is more abundant than foothill paloverde and is not confined to the banks of streamways as it is on the floor of the desert. Velvet mesquite becomes more abundant as a small tree of 6 to 10 feet (2 to 3 m), and saguaro and ocotillo are closely restricted to warm sites. In the rugged area drained by the Gila River, jojoba is strongly dominant on many north slopes. At desert levels here it is not uncommon to find large perennials which are characteristic and abundant at higher elevations, including sotol, burrowweed, rosetree, Wright bushmint, redberry buckthorn, and shrub live oak. Through a vertical range of 1,300 feet (390 m) above the limit of the desert there is a gradual disappearance of characteristic desert plants. Isolated individuals or colonies of certain ones may be found as high as 6,000 feet (2,000 m).

There are over three hundred perennials that either occur infrequently and locally or else are very inconspicuous elements of the vegetation. On the rough slopes, particularly those of northern or eastern aspects, selaginella—clubmoss, sometimes called resurrection plant—forms an almost continuous carpet, gray and inconspicuous in the dry months but a vivid deep green during the summer rains. Several grasses, commonly annuals, often find the conditions favorable enough to live through two or more rainy seasons. Under ledges of rock and in the shade of large trees are found small ferns which appear completely at home in the desert.

The converging ribbons of denser vegetation along the small streamways are very conspicuous when a plain or bajada is viewed from an elevation or from the air. As soon as a drainageway becomes large enough to have interrupted deposits of sand in its bed, the marginal vegetation begins to be more abundant. When the sandy bed becomes nearly continuous and reaches a width of 3 to 6 feet (1 to 2 m) the marginal vegetation becomes still more dense. Creosote bush responds notably in height and bulk to the slight improvement of moisture conditions; velvet mesquite, whitethorn, and catclaw contribute more than any other plants to the increased density. Species found only along the streamways are blue paloverde, Mexican crucillo, lycium, broom baccharis, Thurber anisacanthus, and ambrosia bursage. Several species of *Opuntia,* notably staghorn cholla, cane cholla, desert Christmas cactus, and pencil cholla retain their relative abundance along the streamways.

The alluvial floodplains of the Arizona Upland are among the most favorable desert environments with respect to water supply and depth and texture of soil. The original vegetation was a nearly pure stand of large velvet mesquite trees. These trees long constituted the only source of good fuel in the region, and the soil on which they grew was the most fertile, tillable, and easily irrigated land. Nearly all the alluvial floodplains have now been denuded of trees or have open stands of small second- or third-growth trees that have sprung up from suckers of the original stumps. In some places the rivers' changing regimen has covered the clay bottoms with sand and converted floodplains from the alluvial to a sandy type. In many areas the original woody plants have been replaced with the exotic salt cedar, a plant with high water use and of doubtful value except for wildlife.

## Plains of Sonora

Lying between the foothills and the coastal strip in Central Sonora is an area that has some physical features in common with the Lower Colorado Valley. The general surface is very regular, with an inclination from an elevation of 2,500 feet (750 m) at the inner eastern edge to 330 feet (100 m) on the outer. The mountains and hills are few and do not exceed 10 percent of the total area. Only four mountains rise above 3,300 feet (1000 m). The surface is coarse or fine gravelly outwash, with only small accumulations of sand marking the terminus of clogged arroyos. Although some of the mountains are volcanic, there are no recent malpais fields. The rainfall is 10 inches (250 mm) to 15 inches (375 mm) an amount substantially greater than in the Lower Colorado Valley. Frost is infrequent and less severe than in the areas to the north. Summer temperatures are more moderate than those in the Lower Colorado Valley.

The Plains of Sonora constitute the smallest and least diversified of the vegetational subdivisions of the Sonoran Desert. They constitute the southern termination of the very arid type of continental desert extending north to eastern Oregon, and on the other hand include botanical features that characterize the west coast of Mexico far to the south.

Over most of the area the vegetation is a very open forest of small, low-branching trees, with irregularly placed colonies of shrubs not tall enough to impair the view, and with large but widely spaced columnar cacti. The dominants of the Lower Colorado Valley—creosote bush, white bursage, and triangle leaf bursage—are of local occurrence. Colonies of *Opuntia* are relatively infrequent; there is a distinct approach toward association of plants that are more uniform in habit. In the Plains of Sonora, paloverde, ironwood, and mesquite find their optimum conditions and show their maximum size, abundance, and universality of habitat. In the central and southern parts of the Plains these trees are joined by a number of subdominants, only one of which—copal or torote—extends north of the international boundary. The gradual enrichment of the flora also extends to the shrubs and herbaceous perennials, although not to the ephemerals. In this subdivision, low stature, open spacing, and diversity of life forms that are the cardinal features of the desert are still prevalent but are becoming less noticeable.

Both the vegetation and the flora of the Plains of Sonora are influenced by the transitional position of the area from desert to thorn forest. The gradual enrichment of the flora from north to south is an outstanding feature of the area. There is also a strong contrast between the vegetation of hills and that of plains, and the extension of the newly encountered plants much farther north on the hills than on the plains causes an intermingling of the two principal differentiating features. North of Carbo the vegetation is little affected by the influx of southern plants. Colonies of creosote bush alternate with open forests of foothill and blue paloverde, ironwood, mesquite, ocotillo, and organ pipe cactus; subdominants include several chollas and senita. The vegetation is somewhat thicker on volcanic hills.

In the central region between Carbo and Hermosillo, the vegetation increases in density. Colonies of creosote bush are still frequent, and summer growing annuals and perennials are abundant, making this region appear less barren than areas to the north. Brittlebush becomes larger and several woody plants begin to appear, becoming abundant a little farther south. Trees and shrubs with distinctive characteristics, though not dominant, give the landscape a different aspect. These include MacDougal ocotillo, Sonora rathbun cactus, torote, Sonora paloverde, lignum vitae, cudjoe wood, and tree morning glory. As the density of vegetation increases, organ-pipe cactus, senita, and other species of cactus become less abundant.

In the southern region, south of the Río Sonora, are a number of other additions from the south: creosote bush drops out; brittle bush becomes less abundant, but frequently reaches a height of 6 feet (2 m); and the saguaro reaches its southern limit.

New additions from the south include the large green tree palo San Juan, Willard acacia, boatthorn acacia, Sonoran cordia, pochote, apes earring, and hairbrush cactus.

## Foothills of Sonora

Shreve devoted considerable time to studies of the Foothills of Sonora, although he recognized that it was the least desert part of the Sonoran Desert. The areas with typical desert vegetation usually occur in strips bordered by grassland or forest.

This southeasternmost subdivision of the Sonoran Desert extends from the vicinity of Arizpe, on the Río Sonora, southward to the delta of the Río Yaqui, falling gradually from about 3,300 feet (1,000 m) to sea level. On the west it passes with remarkable sharpness into the Plains of Sonora; on the north it follows up the valleys of Río Sonora, Río Moctezuma, and Río Bavispe; on the east it extends to the oak grassland of the lower mountain slopes; and south of 28°N it merges gradually into thorn forest. Except at the southern end the entire area is rolling or hilly, with only narrow stretches of plain at intervals along the larger rivers. The mineralogical material is chiefly older volcanics or limestone, but there are large lava fields in the Moctezuma Valley and between Río Moctezuma and Río Sonora.

Much of this subdivision's surface is rugged but not mountainous. Two-thirds of the Foothills area lies more than 60 miles (100 km) inland from the Gulf of California. The rainfall of the Sonoran Desert is greatest here, sometimes reaching 20 inches (500 mm); 70 to 80 percent, however, falls in the three summer months, and there is a period of deficient rainfall usually lasting from January or February to July. Though the vegetation of the area is distinctly desert in character, it is heavier and more continuous than in any other of the subdivisions of the Sonoran Desert.

On the north the Foothills area is bounded by a temperate type of grasslands, and on the south by arid subtropical thorn forest. On the west it is sharply limited by the level arid coastal plain, and on the east by the lower slopes of the outlying ranges of the Sierra Madre Occidental, where an increase of rainfall is encountered.

A few of the dominant plants of the Sonoran Desert are more abundant here than elsewhere, but the majority are common in adjacent areas and restricted to the Foothills habitats most nearly resembling their areas of greatest abundance. The adjacent thorn forest has contributed heavily to the flora of the Foothills of Sonora, and many of these plants are conspicuous in the vegetation of the southern half of the area. The great majority of these plants reach their northern limit in the Foothills of Sonora and are absent from the same latitude in the Plains of Sonora and the Central Gulf Coast.

The most noticeable features of the vegetation are the abundance of small trees and the amount of grass at higher elevations, the infrequency of cacti, the occurrence of palms, and the appearance among the dominant perennials of many trees and shrubs not found elsewhere in the Sonoran Desert.

On the floodplains and well-watered level areas, mesquite is the dominant tree in the north, but in the south it shares its dominance with boatthorn acacia and lysiloma. The most common trees are torote, miguelita, Sonora paloverde, ironwood, MacDougal ocotillo, lignum vitae, and pochote. Large cacti, yuccas, and agaves are less frequent in the Foothills and not so conspicuous as in more open vegetation. In general, the vegetation is more monotonous. The columnar hairbrush cactus is a common associate with the other trees.

In the southern Foothills the prevailing vegetation exhibits the characteristics of the desert, but 20 to 30 percent of the area is occupied by littoral or thorn forest vegetation, including mangrove swamps, marshes, and saline flats.

This area marks the transition in which the outlying bodies of the thorn forest occupy the most favorable habitats and the desert vegetation is found in the driest places. The southern edge of the desert is also the northern edge of the tropics. As the valley of the Yaqui River is approached from the north, some of the characteristic features of the desert begin to wane and the vegetation takes on characteristics suggestive of tropical jungles.

The amount of bare ground between plants becomes less, for example, and there are more areas in which there is no bare ground. Trees become taller, and there are more kinds of plants in a single locality. While thorny trees and shrubs with small leaves are still predominant, there are occasional ones with leaves larger than any that are found in the pronounced deserts. The cacti of the open desert rapidly disappear and new shade-enduring species appear, playing a much less important role in the vegetation than do their congeners to the north.

It is obvious that conditions are much more congenial although the region is still an arid one. Nevertheless, there are hundreds of species in the Valley of the Yaqui that have not been able to extend their ranges 100 miles (160 km) farther into the still drier plains of central Sonora.

Although it is not actually part of the Sonoran Desert, the thorn forest is of importance because of its relation to the desert. Shreve (1937b) in a study of the vegetation of Sinaloa, noted that lowlands of the extreme southern end of Sonora and the entire state of Sinaloa are occupied by thorn forest. Here is found much convincing evidence of the derivation of some desert species, the origin of some structural and physiological peculiarities of desert plants, and the beginnings of certain features of vegetational structure which the desert manifests.

## Central Gulf Coast of Sonora

The Gulf Coast of Sonora is occupied in part by hills and small mountains rising abruptly from the beach and in part by bajadas extending far inland. The most mountainous area is north of Guaymas. The soils are shallow, coarse, and covered with rock fragments. Dunes are infrequent and small, and lagoons, marshes, and mangrove swamps are few.

The transition from the dunes and sandy plains of the Lower Colorado Valley to a hard surface takes place about 27 miles (43 km) north of Libertad. Here, creosote bush and white bursage lose their dominance and blue paloverde, ironwood, torote, and ocotillo becomes prominent. Along the coast, plants of saline habitats are most common.

The rainfall of the Central Gulf Coast is very low and uncertain. The summer temperatures are very high, except when those of the immediate shore are ameliorated by onshore winds. The hills are bare and the streamways are bordered by aprons of sand and boulders. The rain is sporadic and may fall either in the latter half of the winter or in midsummer, but is frequently lacking, sometimes for several seasons in succession. Based on the few available records, this is the driest subdivision of the Sonoran Desert with rainfall varying from 4 inches (100 mm) to 10 inches (250 mm).

The largest islands in the Gulf, Angel de la Guarda and Tiburón, as well as numerous smaller islands, are also part of the area. The vegetation of the Central Gulf Coast extends inland for 12 to 18 miles (19 to 29 km) and merges gradually into that of the Lower Colorado Valley or the Plains of Sonora.

On the coast, 6 miles (10 km) south of Libertad, is the only colony of the boojum tree known on the mainland. Boojum trees occupy the north slope of a low granite ridge which runs east from the beach and follows the ridge inland for about 2 miles (3 km). There are none on the south slope of the ridge, and none have been found anywhere else in Sonora. Shreve estimated that the total number of individuals was probably about 2,000. Reproduction is good, and the tallest individuals are 40 to 45 feet (13 to 15 m) high, as compared with the highest known trees, 72 feet (24m) high.

The boojum and the three species of ocotillo that are abundant in the Sonoran Desert have worked out a system of withstanding a dry climate unlike that of the cacti or any other plants. Food material is made by the leaves, as in most plants, and has not been assumed by the stem as in the cacti. The leaves appear only at favorable seasons and are not highly specialized to reduce water loss, as are the leaves of so many desert plants. In short, they are capable of a high food-making performance as long as they last.

When the leaves fall, the boojum and ocotillo are transformed from veritable moist country plants to truly desert ones. The stems are not capable of adjusting their size to the amount of water they contain, as the accordian-pleated trunks of the giant cacti are, but make their adjustment of volume internally. The water that they contain is tightly held by heavy layers of surface cells impregnated with resin.

Boojum is very abundant throughout the central third of Baja California, and its occurrence at a single locality in Sonora is probably due to a chance natural introduction rather than to survival from a former wider occurrence.

Tiburón Island and the neighboring coast are of interest as the southernmost locality in Sonora for spring allthorn, triangle leaf bursage, jojoba, California barrel cactus, pencil cholla, and desert saltbush, as well as the northern limit for American mangrove, smooth colubrina, bushmint, mangle dulce, Willard acacia, and viscainoa. Viscainoa, like the boojum, is abundant in Baja California and is known to exist in Sonora only at three stations between Tiburón Island and Guaymas.

Guaymas is cited as the type locality for a large number of plants, giving rise to the impression that there is a high degree of endemism in the flora of its environs. The Guaymas region was far more accessible than any other locality in central Sonora until about 1900 and was the first place visited by active collectors. Very many of the plants first found at Guaymas were later known to have a wide distribution in central and southern Sonora or even in Sinaloa. There remain, however, perhaps forty species that are not known outside a radius of 32 miles (50 km) around Guaymas.

## Baja California

At the north end of the peninsula, the lofty, uniform crest of Sierra San Pedro Martir terminates abruptly just north of the thirtieth parallel. North of this latitude, Baja California has three longitudinal belts of very dissimilar vegetation: chaparral on the coast and westerly slopes, coniferous forest on the mountains, and desert on the Gulf coast. South of this latitude, as far as the northern end of Sierra de la Giganta, the peninsula is desert from coast to coast. South of Sierra San Pedro Martir the peninsula maintains a fairly uniform width of about 50 miles (80 km) for a distance of 150 miles (240 km). It then widens, with a long triangular projection on the Pacific coast, and has its maximum width of 120 miles (190 km).

Baja California includes two subdivisions of the Sonoran Desert: the Vizcaíno Region, and the Magdalena Region. It also includes parts of two others: the Central Gulf Coast Region and the Lower Colorado Region.

Baja California has become well traveled, owing to the completion of a paved highway from Tijuana to the southern tip of the peninsula. South of El Rosario, 225 miles (360 km), from the border, the highway traverses a portion of the Vizcaíno Region, then turns to the Gulf coast where it follows the Central Coastal Region from Santa Rosalia to Puerto Escondido; heads inland and continues on through the Magdalena Plains and finally crosses the thorn forest south of La Paz on the way to San Lucas.

Travel conditions were vastly different when Shreve made trips to Baja California. He once stated that the best way to explore the area would be with an autogiro—the worst way with an automobile. However, the expeditions of 1934 and 1935 in cars were both made without serious difficulty.

According to Shreve (1935a): "The requisites for such a trip are some previous experience with similar country, adequate preparation and equipment, total disregard for varnish, fenders, rubber and other superficial features of the vehicles, and an unlimited amount of patience."

In 1934 there had been no rain for several years and the sandy or alluvial stretches of the road were rather heavy. In 1935 there had been rain previous to the start of the expedition and it continued to rain for two weeks after the expedition left the border. Although total rainfall is low in desert areas, the few rainy periods seem to occur at the most inopportune times for the traveler. Camping is made damp and troublesome. Streams are encountered where dry crossings were expected. Roads that are usually good become almost impassable: the fine clay soils which normally make a firm surface are converted into bottomless muds of the most tenacious consistency and are covered with water of unguessable depth.

The 1935 expeditions included Shreve, Wiggins, Mallery (sometimes referred to as Dwight or T. D.), and Jack Whitehead.

*On a trip into Baja California (February–March 1935) we reached Hamilton's Ranch late on a Saturday afternoon and decided to rest up over the week-end and fill up on Hattie Hamilton's good food. Monday morning it was threatening rain and Mallery was in a dither to get as far south as possible before the rain began, for Forrest has forgotten to put in tire chains and T. D. was afraid we'd get badly stuck in the mud of the San Quintin Plains if rain fell while we were in that area. For some reason Forrest decided to rearrange the gear in his car before hitting the road. Frankly, I'm sure he thought the threatening clouds would blow over, but they did not and his car got stuck on the first hill south of Hamilton Ranch (about a mile south) because the rain did hit us just after we drove out the ranch gate.*

*We had a rugged day of it, and were almost continuously digging the car out of a boggy mudhole, cutting and carrying brush, pushing and straining to make any progress at all with the ponderous old Willys-Knight. We camped under a group of Monterrey Cypress trees just 19 miles [30.4 km] from Hamilton's Ranch shortly after dark that night. A drizzle continued throughout the night but stopped about daybreak. By 10 A.M. we had partially dried our soggy bedrolls and were again fighting mud. At 2 that afternoon we were two miles [3 km] from where we had camped the night before.*

*But then we got a break. We got out of the alluvial silt of the llano and made fair time over gravelly and sandy soil until we reached Rosario a bit after dark. We were hospitably allowed by Sr. and Sra. Espinosa to occupy one room of their three-room casa in order to get out of the rain that had begun to fall steadily. The roof leaked, but we managed to arrange our sleeping bags partially under the cots on which Shreve and Mallery slept so all four of us managed to keep fairly dry that night. It cleared again shortly after daybreak, but started to rain again in the late afternoon. Another night in the crowded quarters of Espinosa's establishment, and another, and a fifth one. Finally, on the sixth day it stayed rainless until about 3 P.M. and we made it to Quail Springs 19 miles [30.4 km] from Rosario before dark.*

*An early start the next morning got us ahead of the rain which appeared to follow us a few miles in the rear all day. We reached El Marmol at 11 P.M. and were given shelter at the onyx mine by Mr. George Brown, who was superintendent at that time. We stayed there for five days, with rain falling most or all of the time at night and during parts of each day, and finally got away on the morning of the sixth day . . . and I suspect that Mallery still curses the chainless Willys-Knight once in a while. [Wiggins, personal communication]*

The troubles for these fellows did not end here but conditions did improve as the party moved southward. Highlights of the first days of the trip come from Shreve's diary.

### TUESDAY, FEB. 5, 1935
*We leave camp at 10:45 and fight mud for 5 miles [8 km] cutting brush, looking for good detours and jacking up buried wheels. When we get out of the flats the going is better but we still have delays and road trouble to the top of the mesa. The descent into Rosario is good and about 8 we stop at Sr. Mesa's store and he gives us a room for the night and Sra. Mesa makes some good coffee. 36 miles to-day [57.6 km].*

### WEDNESDAY, FEB. 6, 1935
*There was rain in the night. The leaky end of our room was provided with lard cans to catch the drip.*

### FRIDAY FEB. 8, 1935
*The morning is cold and foggy, but sunshine follows and we decide to push on. We find that Sr. Mesa hasn't as much gas as we need, which is not pleasing. We find the road dry in general but have trouble with a few wet spots and Dwight gets hopelessly stuck. Two Mexicans in the saddle help us through, and we soon leave the Rosario Valley for gravelly branch of it and the going is good to the old adobe, where we camp.*

### SATURDAY, FEB. 9, 1935

*It clouded up in the night and began to rain about daybreak. We made a hasty breakfast to get started before the red hills get slippery. We get over the top of the Sierra Aguajita (about 2000 ft.) [600 m] all right but get stuck badly three or four times on the clay flats, with showers every now and then. This clay mud is very adhesive to feet, shovels and everything. Down on the gray soil the going is very good. We made 10 miles [16 km] from 9 to 3, but push on 52 miles [83km] more on fair roads to El Marmol. Mr. Brown gives us fine quarters in a palm thatched adobe.*

### SUNDAY, FEB. 10, 1935

*As usual, it rained in the night. We got up late, had flap-jacks, bacon and eggs and spent the day drying out, cleaning up and trying to remove the tenacious red clay from everything. It is windy and cold. The weather and vegetation seem more like New Mexico than B.C.*

### MONDAY, FEB. 11, 1935

*The night was cold (43°) [6.1°C] and there were showers in the early morning. We all do various odd jobs. I go out in a nearby hill to take a plate photo. Wiggins and Jack climbed on the mesa east of here from which they saw the Gulf.*

### TUESDAY, FEB. 12, 1935

*Still at El Marmol. . . . At 10 A.M. it was 46° [7.8°C] and the maximum for the day was 52° [11.1°C]. We all spend most of the day in the kitchen trying to keep warm. . . . We are determined to stay right here until it clears up and gets warmer.*

### THURSDAY, FEB. 14, 1935

*Still at El Marmol. In default of burros, Wiggins, Jack W. and I start at 10:30 up the trail toward Miramar. We see veritable forests of Pachycormus.*

### SATURDAY, FEB. 16, 1935

*We have a head wind up the long grade to the summit of the B.C. road—2800 ft [848 m]. At the lunch stop I try some photography in spite of the wind. We find the Laguna Chapala wet and take the wet-weather road around the E. end. We camp in sight of the Lake sheltered by granite hills. The wind is not so bad as it was.*

### SUNDAY, FEB. 17, 1935

*We cross a granite ridge with magnificent vegetation and a fine view of the Sierra San Borjas. We cross a stretch of sand and then enter the long open forest of Pachycormus leading down the valley to Punta Prieta. We camp 3 miles [5 km] N. of P.P. in an ideal and beautiful spot. No clouds, no wind, much warmer, a full moon. We all feel cheerful and elated. Only 34 miles [54 km] to-day but we collect lots of plants, notes and photos.*

Challenged by the rains and winds of unpredictable arrival and intensity, and covering only about 20 to 30 miles (32 to 48 km) daily—sometimes 0—the expedition traveled for more than two months and 1700 miles (2,720 km).

## Vizcaíno Region

The Vizcaíno Region comprises the Pacific drainage of central Baja California from the southern end of Sierra San Pedro Martir south nearly to latitude 26°N. The designation of this subdivision of the Sonoran desert is an extension of the name "Vizcaíno Desert," which has long been applied to the plains of the long triangular projection. This region includes the rough mountainous interior as well as the Vizcaíno Desert and an additional wedge of coastal plain that extends south to Punta Pequeño.

The Vizcaíno plain is the largest area of Baja California in which a nearly uniform vegetation extends for many miles in every direction. Along its inner edge, the plain is entered by a few streamways, all of which spread and disappear in a short distance. The behavior of local drainage is indicated by bands of heavy vegetation separated by broader bands of nearly bare desert pavement at almost the same level. A few dunes were found in the northernmost part of the plain near the last of the coastal hills. All the dunes were very active, and their only plants were in thickets at the bases of lateral slopes. The entire Vizcaíno plain lies below 330 feet (100 m) elevation, and its inner edge is sharply marked by a sudden rise onto volcanic mesas.

At the time of Shreve's descriptions, very little was known about the climatic condition of the Vizcaíno Region as there were no official records available. The rainfall in the Vizcaíno Region occurred almost wholly in the winter and early spring. At long intervals there were summer storms south of the Vizcaíno plain. According to the inhabitants, the winter rainfall is extremely variable from year to year. In 1934 the peninsula was very dry, and residents stated that it had been seven years since good rains. During the expedition of 1935, heavy rains were encountered in February, and the condition of the plants indicated that rain had begun several weeks before. Inhabitants alleged that the rains had not been so copious for twelve years. Whatever the conditions may be from January to April, the remaining eight months of the year are almost invariably dry through the Vizcaíno Region. During spring and summer there is morning fog along the coast as far south as the Vizcaíno plain, and it extends inland wherever there is a low coastal shelf for a distance of 3 to 4 miles (5 to 6 km). Foggy mornings are said to be most frequent in July and August. A strong wind from the Pacific whips the coast during the day and early part of the night, and is apparently continuous throughout the year.

Shreve noted that the greatest contrast in the area was between the mountainous axis of the peninsula and the alluvial deposits of the Vizcaíno plain. Proximity to the sea means little with respect to rainfall, humidity, or soil moisture, but the extreme coastal belt is invariably open and barren because of the incessant high wind.

Under comparable conditions of substratum there is a difference in vegetation between slopes facing north and south. At higher elevations the difference in composition on opposed slopes is great. On hills lying within 2 to 3 miles (4 to 5 km) of the coast there is a marked difference in vegetation between the slopes facing the sea and those facing inland. The windward slopes are usually without any large perennials, may be scantily covered with low perennials, or may be so bare that the yellow lichens covering the rocks color the entire slope, conspicuous for several miles.

Traveling south, the observer sees gradual but continuous changes in the plant life. The landscape is dominated by a few large species usually greatly outnumbered by some of their smaller associates. A strong individuality is given the vegetation of the southern Vizcaíno Region by the boojum tree, elephant tree, agave, cardón, and datillo. Although these five perennials never total more than 20 percent of the population, they give character to the landscape. On the immediate coast and in some of its least favorable inland habitats, the elephant tree is a prostrate and grotesquely formed tree with a very thick trunk narrowing abruptly into normal slender terminal branches. Over most of its inland range, however, it is an erect and shapely tree. The basal diameter of the trunk is always greatly exaggerated and it tapers rapidly among the branches. The smooth, light bark contrasts strongly with the dark foliage to make it a highly ornamental tree. The dominant plant of the Vizcaíno plain is the datillo, which forms from 60 to 80 percent of the number of large perennials throughout the northern part of the plain. The tallest individuals are from 23 to 33 feet (7 to 10 m) high. There is no suggestion of a foliage canopy among the trees near the observer; however, when an expanse of datillo forest is seen from a slight elevation, it gives the impression of an open canopy not unlike that of the dry pine barrens of the southeastern United States. Datillo grows chiefly in close groups of five to fifteen plants. It is erect until it reaches about half of its mature height; then it begins to lean. Near the coast the trees are inclined away from the wind, but in the interior they bend grotesquely in every direction. Very rarely do the individuals or groups of datillo have the symmetrical and balanced form that is so obvious in the Joshua tree in the Mojave Desert.

The vegetation of the Vizcaíno Region comprises relatively heavy stands of large perennials, in communities of rich composition. It also displays every phase of the impoverishment of these outstanding communities—some of these stands being vegetatively the most important. Finally, there are great saline areas in the Vizcaíno plain on which there is an extremely scanty cover of halophytic plants.

There are many leaf succulent plants in the Vizcaíno plain as indicated by the fact that twenty-three species of *Agave* have been credited to Baja California. And, according to Shreve, the ecological prominence of the genus in this area is matched only by the extensive stands of lechuguilla in Coahuila.

## Magdalena Region

The Magdalena Region is the southernmost part of the Sonoran Desert, extending 200 miles (320 km) farther south than the mainland part. The eastern boundary is formed by a continuous series of higher mountains than in the Vizcaíno

Region. The northern half of the region, between San Ignacio and Comondu, is an intricate field of volcanic hills and mesas, with a narrow, sandy coastal plain. Much of the area is occupied by long stretches of volcanic eruptives falling at a uniform gradient from the central series of extinct cones. The steepsided volcanic tongues are of predominantly uniform level in transverse section, with a rocky surface and shallow soil. Between them are narrow barrancas with soil and local seepages of water.

In the northern Magdalena Region, north of Comondu, volcanic mesas and slopes form the prevailing surface. There are extensive malpais fields with scattered pockets of soil, and also areas of thin but continuous soil strewn with small stones. The vegetation of the malpais and the thin soil of the mesas is open, irregular, and xeric, closely resembling that east of the Vizcaíno plain.

South of Purisima and around the north end of the Magdalena Plain the rainfall is evidently very low. The vegetation is extremely open and almost wholly made up of species characteristic of the very dry areas of Baja California and Sonora.

The landscapes of the Magdalena Region are less striking than those of the Vizcaíno Region. This difference is due primarily to the absence of boojum and elephant trees, and the infrequency and poor development of cardón and all the species of yucca and agave. Also there is a greater uniformity in the life forms, with small leaf trees and shrubs being dominant. On the volcanic mesas there are few cacti, and tall forms are especially infrequent. The chollas are very irregular in occurrence, forming open colonies in one spot and being absent over large areas. Prickly pears, very uncommon in the Vizcaíno Region, are abundant in many parts of the Magdalena Region.

The plain from San Domingo to Venencio, a distance of 112 miles (180 km), is uniform in physical features and monotonous in vegetation. The profile of the vegetation is irregular, being formed by either mesquite or lycium and by projecting tops of cardón. There are no tall cardóns but many massive ones which have branched just above the ground. Other cacti, such as organ-pipe and senita and the cholla, are abundant but are still outnumbered by the small leaf trees and shrubs.

The vegetation of the southernmost tip of the Magdalena Region is dense, low, and poor in perennials. Datillo is prominent in some places, particularly in depressions, but is only 3 to 10 feet (1 to 3 m) in height. Cardón is rare. The peninsular ocotillo is abundant but rarely more than 8 feet (2.4 m) in height. The prevailing shrubs are lycium, brittlebush, jojoba, and caper. East from the inner edge of Magdalena Plain, the slopes of Sierra de la Giganta change rapidly from desert into the Cape thorn forest. Small trees rapidly become more abundant, and, as many of the shrubs are over 6 feet (2 m) high, it is not possible to see very far through the vegetation. Cacti, yuccas, and ocotillo are still frequent, and the trees are nearly all small leaf deciduous types.

## Cape Region

During the 1935 trip to the tip of the Baja peninsula, Shreve observed the southern limit of desert plants and types of plant communities in order to determine the character of the vegetation of the Cape Region and to compare the southern edge of the desert with its termination in southern Arizona.

He noted that the southern end of the peninsula is "biologically" an island because of its effective isolation from the nearest areas of similar climate. It is separated from the Mexican mainland by the Gulf of California, averaging 90 miles (144 km) in width, and from the mesic highlands of northern Baja California by an arid stretch of 340 miles (544 km).

The Cape Region not only has a higher rainfall than the central section of the peninsula but receives it in the most favorable season. To these facts must be attributed the termination or localization of desert and the existence in the Cape Region of vegetation with greater moisture requirements. The plant life, developed under conditions of both geographic and climatic isolation, has undergone little change during most of the period in which the angiosperms have dominated the vegetation of earth.

The extreme tip of Baja California is believed to the earliest part to emerge from the sea, and is full of biological interest. There is a central range of mountains reaching an altitude of

7,300 feet (2,190 m) and bearing small areas of forest. The slopes and plains surrounding the highest mountains are covered with vegetation much denser and somewhat taller than vegetation only 100 miles (160 km) farther north. On the east side of the cape, plants are lower and more widely spaced and cacti are more abundant.

In short, this region is on the border between the desert country, which extends north for 1,600 miles (2,560 km), and the dry thorn forest, extending nearly as far south along the coast of the mainland. This change comes, in fact, a little farther north in Sonora than it does in Baja California.

It is interesting to note in the transition region on both sides of the Gulf that the thorn forest plants grow in localities with a fair supply of soil moisture and the desert plants in the drier situations. It often happens that a slight change in the character of the soil modifies the moisture conditions and makes as much difference in the plant life as a jump of 200 miles (320 km) in latitude.

Although a plant is unable to move, its race is constantly moving by the dispersal of seeds. In unfavorable localities the seedlings perish; in favorable ones they live and grow to shed more seeds in the same spot. In this way every plant race is constantly testing the possibilities of new locations. This behavior is strongly exhibited in the transition region between desert and thorn forest, where pioneers of one region are actually passing the pioneers of the other and settling far behind them.

The Cape forest below 3,300 feet (1000 m) is distinctly xeric. The canopy of the forest is usually open and always extremely irregular. Shrubs are almost invariably abundant and in slightly moist situations contribute to the formation of impenetrable thickets. One of the commonest shrubs is yellow trumpet with a height and stoutness of stem that almost give it the rank of a tree. Several shrubs belonging to the sunflower family equal the trees in height, and coral vine is abundant in all but the driest situations. Its clusters of crimson flowers give much vivid color to a floral display in which yellow is predominant.

The relatively rich composition of the Cape forest; the close mingling of trees of different height and branching habit;

the occurrence of shrubs, erect and compact, broad and poorly branched or semiscandent; the presence of cacti, yuccas, and vines give the Cape Region the appearance of impoverished tropical jungle.

The vegetation and growth of the Cape Region bears some resemblance to the thorn forest of Sinaloa in height and density. The Cape forest extends from sea level to the borders of mountain type of forest found above 3,300 feet (1,000 m). A short distance north of Todos Santos it gives way to low desert scrub which characterizes the outer edge of the Magdalena Plain. On the Gulf Coast it covers all of the granitic outwash and certain favorably located areas of volcanic outwash but is not found on the volcanic hills near the coast south of La Paz. In the Sierra Giganta it occupies the slopes of the mountain on the west and the upper slopes on the arid eastern side.

The number of plants common in the Cape Region but absent from the desert of Baja California is large. The infiltration of the Cape vegetation by the number of desert plant species is considerable, but only locally important with respect to their role in the vegetation. Creosote bush reaches the top of the southern end of the escarpment but does not descend into the plain of La Paz. Cardón, organ-pipe cactus, paloverde, torote, and other plants prominent in the desert are also frequent in Cape vegetation. Many of the common plants of the desert are abundant on the volcanic areas of the Cape Region but only sparingly represented in the Cape forest.

# Desert Mountains

THE SOUTHERN HALF OF ARIZONA is characterized as a relatively level plain studded with numerous hills and mountains. The plain rises from elevations of a few hundred feet (a hundred m) along the Colorado River to as much as 4,500 feet (1,350 m) and 5,000 feet (1,500 m) near the New Mexico boundary. The lower elevations follow the Gila, Salt, San Pedro and other rivers, while the higher plains surround the loftier mountains of the southeastern portion of the state. Between the Colorado River and Tucson, there are no mountains of commanding elevation, and the area occupied by the scattered volcanic peaks and ranges is not more than one-tenth of the total area of the region. Geologically, this region has remained unchanged throughout a long period and the mountains and hills have been subjected to prolonged erosion, serving to build up the shelving plains which form the intervening valleys.

The general character of the vertical zonation of plant and animal life on the desert mountains of the Southwest is similar in all of the ranges from the Colorado River eastward to the Pecos. Rising from plains with desert, desert-grassland transition, grassland transition or grassland vegetation, the altitudinal zones, shown schematically in Figure 4.1, may be roughly grouped as desert, encinal, and forest (chiefly coniferous). The elevations at which encinal or forest will be encountered in ascending a desert mountain will depend primarily on whether observation is being made of a ridge or valley, or of a north or south slope. The lowest occurrence in valleys is an extremely variable item, as individual forest trees will sometimes descend along streams 3,000 feet (900 m) below their lowest north slope occurrence.

If the modifying influence of topographic irregularity is disregarded, the lowest normal occurrence of encinal or forest will be found to depend on three factors: (1) the elevation of the plain from which the mountain rises, (2) the total elevation that the mountain attains, (3) the mineralogical character of the mountain and consequent nature of its soil.

Studies conducted by the Desert Laboratory staff were not limited to the deserts but extended up into the mountain areas as well. Here, the uniformity of desert vegetation was broken: in the higher lands or lower mountain zone with juniper and oaks; in the higher mountain zones with various pines and cedars, maples, aspen; and in the highest mountain zone with the Alpine formation. In the Santa Catalina Mountains over as short a distance as 20 miles (32 km), a vertical zone is

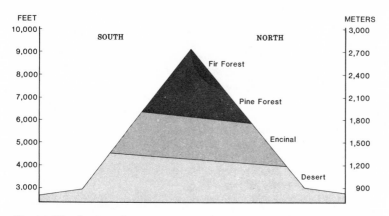

Fig. 4.1. Distribution of major vegetation types in the Santa Catalina Mountains (after Shreve, 1915).

crossed in which plants represent all of the temperature-related plant zones in North America with the exception of the tropics and the extreme north.

In 1915, Shreve wrote a comprehensive report on vegetation related to environment, entitled *Vegetation of a Desert Mountain Range as Conditioned by Climatic Factors.* This paper was one of Shreve's major contributions in his comprehensive coverage of the vegetation of the southwestern United States and northwestern Mexico. He discussed the general geography and topography, the vegetation, the phytogeographic relationships of the flora, the climate, and the correlation of vegetation and climate of the Santa Catalina Mountains. Since the original publication is out of print and is available only in a few libraries, it is reviewed in detail here.

*The desert mountain ranges of the southwestern United States stand in the midst of a region which presents severe conditions for plants. The relative richness of the vegetation in this region is due chiefly to the occurrence of two yearly seasons of rainfall. The entire annual vegetational behavior is related primarily to the moisture seasons and much less pronouncedly to the thermal seasons. The perennial plants lead an existence which permits of rapid growth during the warm humid season, together with an extremely low ebb of activity during the arid seasons, and with the possible loss through drought-death of much of the growth that has just taken place.*

*The severe conditions of the desert environment cause the vegetation to exhibit a high degree of sensitiveness to slight topographic and edaphic differences. Wherever the character of the soil or the topographic location is such as to present a degree of soil moisture slightly above that of the general surroundings, or as to maintain it for a longer time in the periods of extreme aridity; or in whatever locations plants are protected from the most extreme conditions of transpiration—in such places are to be found heavier stands of vegetation or else particular species of plants.*

*The higher mountains of the desert region exhibit strong gradients of change in climate and in vegetation. Both of these gradients are much more pronounced than those of mountains of equal elevation in more humid regions. They lead from arid to humid, or at least semi-humid, conditions of moisture, and from sub-tropical to temperate conditions of temperature; from low, open microphyllous and succulent desert, through a sclerophyllous semi-forest to heavy coniferous forest. [Shreve 1915]*

The Santa Catalina Mountains, one of the most westerly of the high ranges of southeastern Arizona, rise from an approximate basal elevation of 3,000 feet (900 m) to a height of around 9,150 feet (2,745 m). Their vegetation is typical of a large number of mountains in Arizona, southern New Mexico, and northern Mexico.

Shreve's study was confined to the south side of the Santa Catalina Mountains, the southern face of which is made up solely of gneiss in varying degrees of hardness. The main ridge and the northern and eastern lateral ridges are worn into a relatively rounded topography, while the southwestern corner of the range possesses rock of greater durability and is correspondingly more rugged in topography.

Shreve described the topography and the most important features of Mount Lemmon, which has an elevation of 9,150 feet (2,745 m). There are several streams draining the south slopes known as Sabino Canyon, Bear Canyon, and Soldier Canyon. Although Shreve made this study the hard way—on foot, on horseback, and with pack train—the modern traveler driving along a paved highway can see nearly everything that is mentioned without getting out of his automobile. If he is willing to expend a little more time, he can park at facilities where the U.S. Forest Service provides information on the vegetation. The observer also has available at 1,000-foot (300-m) intervals roadside markers that provide elevation information.

*The journey from the base to the summit of the Santa Catalina Mountains brings to the eyes of the observer a constantly changing panorama of vegetation. New types of plants are constantly being encountered with increase of altitude, while types already familiar are being left behind. There is no portion of the mountain, at least below 7,500 feet, in which a climb of 500 feet does not materially alter the physiognomy of the surrounding vegetation. The course of the vegetational panorama is not merely a gradual transition from the open desert of succulents and microphylls to the heavy fir forest which occupies the summit of Mount Lemmon. There are interposed between these vegetations two distinct belts of plant life through which this tremendous transition takes place.*

*The arborescent cacti and the trees and shrubs of the desert give way gradually to evergreen oaks, leaf-succulents, sclerophyllous shrubs, and perennial grasses. This open but arborescent vegetation reaches a full development and then gives way to pine forest, with a distinctive accompanying carpet of herbaceous perennials. The pine forest is then, in turn, invaded by spruce and fir and the heavy stands of these trees are accompanied by still another assemblage of shrubs and herbaceous plants. The striking character of these gradations of vegetation is not due solely to the contrast between the varied vegetation of the open desert and the monotony of the closed coniferous forests, but is quite as largely due to the striking types of plants which are to be found both in the desert and in the region of evergreen oaks. [Shreve 1915]*

The detailed descriptions of vegetation that follow apply only to the south face of the Santa Catalina Mountains. Limiting his studies to the south slope enabled Shreve to carry out a study of climatic influences over a vertical gradient of 6,000 feet (1,800 m) with soils developed from the same parent materials. The east and west ridges of the south face furnished north and south slopes at all elevations.

Mount Lemmon is characterized by desert on its four slopes, by open evergreen forests or encinal at middle elevation, by pine forests above 7,000 feet (2100 m), and by fir forests on the highest peaks. The distribution of nearly all species of plants is so definite with the relation to altitude and habitat that it indicates they are controlled in their movements and establishment by the operation of physical factors. In contrast with some of the other workers at the laboratory, Shreve stated that the minor influences of slope exposure and other topographic features caused local departures from the normal altitudinal gradient of vegetation, but these departures merely bring a given type of vegetation to an altitude higher or lower than that in which it is commonly found.

The rainfall for forests at higher elevations is about 2½ times as great as it is on the desert, and the soil moisture in the driest portion of the year from five to fifteen times as great, depending on the slope and exposure. Evaporation is three to four times as great on the desert as on the summit of the mountain. Daily seasonal temperatures are approximately 30°F (17°C) lower on the summit of the mountain than on the desert, and the frostless season is half as long in the former as in the latter locality.

Shreve believed that the distribution limits of mountain plants at the edge of the desert were related to the ratio of evaporation to soil moisture in the Santa Catalina Mountains. The upward limitations of desert plants, on the other hand, appear to be due to the operation of winter temperature conditions. Study of the vertical temperature gradient was complicated by the operation of pronounced cold air drainage throughout the slightly forested or unforested portions of the mountain. It was found that the difference between the minimum temperatures on the floor of the canyon at 6,000 feet (1,800 m) and on the summit of a ridge on the same elevation was as great as the normal difference between two stations of the same topographic site nearly 3,500 vertical feet (1,050 m) apart (Shreve, 1915).

Comparisons of climatic gradients of the Santa Catalina Mountains and gradients derived from weather bureau stations situated at different elevations in the valleys of the adjacent region showed that with respect to rainfall conditions the isolated mountains have a greater rainfall at 4,000 and 5,000 feet (1,200 and 1,500 m) than at localities in the valleys at the same elevations. Also that the coldest temperatures of winter at 8,000 to 9,000 feet (2,400 to 2,700 m) are much milder on an isolated mountain surrounded by desert than they are at the same elevation on extensive plateaus.

In the Cañada del Oro and the Sabino, Bear, and Soldier canyons, as well as a few of the larger canyons of the north slopes, Shreve found water at all times of the year in localities where the local configuration of the valley or the occurrence of resistant dikes of rocks forced the underflow to the surface. During the rainy season, water may be found in any of the large drainageways and the heavy local showers of summer often convert even the smallest streamways into rushing torrents for a few hours. Shreve noted that the small size of the Santa Catalina Mountains together with their elevation results in a steep gradient for all of the major streams. There are no parks or mountain meadows such as are present in some of the larger southwestern mountains. The floodplains and alluvial bottoms are all small and scattered. The Santa Catalina Mountains have been worked so far by erosion and weathering that they possess almost no relatively level areas or regions of indeterminate drainage. The precipitous topography gives little opportunity for the accumulation of soil, and it is thin in almost all localities. Throughout the lower portions of the range, below the pine forest, the soil appears to be extremely coarse by reason of the surface coating of angular fragments from 0.039 to 0.19 inch (1 to 5 mm) in diameter. But beneath this coating there is often a fine soil, still mingled with coarse particles but held in place by the mulch of stones, analogous to "desert pavement." The outcropping rock and larger boulders serve to retard erosion and to preserve a soil sufficiently deep for shrubs and trees to find root. There are many deep soil-filled crevices through which the roots of trees are able to penetrate to bodies of deep-seated soil of favorable moisture content.

Above 7,500 feet (2,250 m) the amount of humus, as well as the amount of surface litter, increases with the density of the stand of pines. On the north-facing slopes, clothed with fir forest, the soil is not much deeper than in the heavy stands of pine but is notably richer in organic matter. In the lower mountain region the restricted areas of alluvial soil are fine sand or sandy loam and possess considerable humus, in contrast with the soils of the slopes.

From the base to the summit of the Santa Catalina Mountains there is a constantly changing panorama of vegetation. New types of plants are constantly encountered with increase in altitude, while other types are being left behind. The arborescent cacti, trees, and shrubs of the desert give way gradually to evergreen oaks, leaf succulents, hard leaf shrubs, and perennial grasses. This open shrubby or treelike vegetation reaches full development and then gives way to pine forest with a distinctive accompanying carpet of herbaceous perennials. The pine forest is, in turn, invaded by spruce and fir, heavy stands of which are accompanied by still another assemblage of shrubs and herbaceous plants.

Shreve divided the Santa Catalina mountain vegetation into the following types:

Desert Region
    Upper Bajadas*
    Desert Arroyos and Canyons
    Lower Desert Slopes
    Upper Desert Slopes

Encinal Region
    Lower Encinal
    Upper Encinal
Forest Region
    Pine Forest
    Fir Forest

*The bajadas, a term applied to the outwash material at the base of the mountain, are not truly a part of the Santa Catalina Mountains but are included because of their strong relationship to the vegetation of the mountain slope.

# Desert Region

Under Desert Region Shreve included all those portions of the Santa Catalina Mountains in which the vegetation is open, low, and diversified in the assemblage of growth forms, with a predominance of small leaf trees and shrubs and an abundance of cacti. This type of vegetation covers the upper bajadas and extends up the slopes of the mountain to elevations of 4,000 to 4,500 feet (1,200 to 1,350 m) according to slope exposure. Desert slope vegetation closely resembles the bajada and then loses most of the larger bajada plants with the entry of dominant plants of the encinal region.

The lower bajadas are covered by a vegetation in which creosote bush is always the predominant plant and is often almost the sole plant of more than 2 feet (60 cm) in height over an area extending many square miles.

## Upper Bajadas

The shorter and steeper upper bajadas fringing the southern and southwestern edge of the Santa Catalina Mountains show much more diversified vegetation. Shreve believed that the absence of caliche from the soil was responsible for this.

The upper bajadas along the south faces of the mountains, at about 3,000-foot (900-m) elevation, bear what may be regarded as the most highly developed type of desert vegetation to be found in southern Arizona or northern Sonora. Here are found a greater number of species of perennial plants than in any other distinctly desert situation. On the upper bajadas there are often, in almost equal mixture, from fifteen to twenty-five perennial species of plants of such size as to dominate the physiognomy of the vegetation. The number of individual perennial plants per unit area is greater than in any areas outside the floodplains of such rivers as the Santa Cruz and the Gila. The only areas that compare with the upper bajadas in these respects are the volcanic hills in which basaltic rock has weathered to a fine clay very retentive of moisture, as is well exemplified in Tumamoc Hill, the site of the Desert Laboratory. Ephemeral plants may form a dense carpet over both the upper and lower bajadas in seasons of well-distributed and copious rainfall. In a climb on foot of two hours, from the base of the Santa Catalina Mountains, greater changes in vegetation and flora will be discovered than can be encountered in the 150 miles (242 km) between Tucson and Adair Bay near Puerto Peñasco in Sonora, Mexico.

## Desert Arroyos and Canyons

In crossing the upper bajadas, it is often possible to detect the approach to a very shallow drainageway through which water runs for only a few hours after the severest summer rains. The larger arroyos are more conspicuous by reason of the heavier stand of vegetation along their margins, and in the largest canyons is found the culmination of the influence of surface streams and underflows for the support of vegetation. Although there is only a negligible and short-lived underflow in these smallest arroyos, they present slightly more favorable conditions with respect to soil moisture; the effect of the rainy season is slightly prolonged in them, while the periods of drought are correspondingly shortened. In the larger arroyos there may not be a constant underflow, but there is at least a relatively high percentage of soil moisture, sufficient to greatly reduce the influence of the arid periods upon their plants. In the largest arroyos and in the mountain canyons themselves, there is either a constant underflow, maintaining high moisture content in the soil of the banks and bed of the arroyo, or else there is constant water, either running or standing in pools.

The smallest arroyos have a closer stand of the same plants that can be observed throughout the bajada—notably mesquite, catclaw, and hackberry. Along somewhat larger arroyos are found heavier stands of the above species, together with blue paloverde, hackberry, and other small shrubs and perennial plants.

In arroyos large enough to have a heavy flow of storm water but not large enough to have open pools of water, there may be found several additional species of plants that also occur on the sandy floodplains of the largest canyons. Prominent among these are desert willow, which may attain the size of a small tree, several stream-side shrubs, and some half-shrubs and plants such as the saguaro. In the larger canyons are to be found communities of more mesophytic plants, stream-side and aquatic plants that stand strongly in contrast with the predominant forms of the bajadas.

## Lower Desert Slopes

Upon leaving the uppermost edge of the bajada and com-mencing the ascent of the mountain over abrupt slopes which lie between the larger canyons, a region is entered in which the physical conditions differ from those of the bajada chiefly in the pronounced slope exposure to the south, southwest, or southeast. Here one encounters large masses of rock *in situ,* with the coarse soil limited to small benches, pockets, and fissures. The vegetation of these lowest slopes is very similar to that of the upper bajadas and is composed of a nearly identi-cal flora. Mesquite, foothill paloverde, and catclaw are repre-sented by smaller and less frequent individuals, and both the cylindropuntias (cholla) and platyopuntias (prickly pears) oc-cur somewhat less frequently. Saguaro is even more abundant on the slopes than on the bajadas, being represented by smaller individuals, among which relatively few have reached the size at which branching begins. For saguaro and the above-men-tioned trees, the relatively rapid erosion of the soft gneiss and the shifting of the shallow soil are apparently too great to permit the attainment of great size or age. Ocotillo, brittlebush, and larchleaf goldenweed are even more abundant on the slopes than on the upper bajada, and teddy bear cactus, the most densely spiny of all cylindropuntias, is found exclusively on southerly slopes and cliffs—in very rocky substratum—at elevations below 3,500 feet (1,050 meters). Ironwood and creo-sote bush have not been detected on the mountain slopes.

The summer and winter ephemerals of the bajadas are nearly all to be found on the desert slopes of the mountain but rarely in such abundance as they attain on level ground. The ascent from 3,500 to 4,000 feet (1,050 to 1,200 m) reveals the first essential changes in the vegetation. At the higher elevation nearly all of the typical desert forms may be found, but cholla has become infrequent and saguaro, barrel cactus, and ocotillo are conspicuously confined to southerly slopes. The blue palo-verde, confined to arroyos in the bajada, is found here growing with staghorn cholla, which attains a maximum height of 12 feet (4 m), and jumping cholla. Engelmann prickly pear and Toumey prickly pear are abundant representatives of the platyopuntia group. The evergreen creosote bush is greatly outnumbered by ocotillo. The globular barrel cactus, or bisnaga, attains a height of 4 feet (1.2 m) with an even greater girth. Similar in form but never exceeding a foot (0.3 m) in height are hedgehog cactus and fishhook mammillaria. The evergreen jojoba and the relatively large-leaved deciduous sangre de drago are frequent, while a large number of less striking shrubs are common including triangle leaf bursage, burroweed, brittlebush, gray thorn, lycium, hackberry, krameria, trixis, desert zinnia, and paper flower.

Summer and winter rains bring two wholly distinct sets of herbaceous ephemeral plants, at the same time that each season causes activity in the root perennials. With less favorable conditions, mesquite, which eventually exceeds littleleaf paloverde in vertical distribution by nearly 500 feet (150 m), is more abundant at 4,000 feet (1,200 m) than it is on the lowest slopes and becomes much larger at 4,200 feet (1,260 m) within 600 vertical feet (180 m) of its upper limit. Such common shrubs of the bajada as squaw bush, krameria, sangre de drago, and hackberry are now very sporadic in their occurrence, and the vividly green larchleaf goldenweed has become very frequent and conspicuous, together with sagebrush and the shrubby buckwheat. On northerly slopes, just below 4,000 feet (1,200 m) are encountered the first individuals of rosetree, sotol, and Palmer agave. Along the arroyos the most conspicuous forms are coral bean and wild cotton.

## Upper Desert Slopes

The slopes lying between 4,000 and 4,500 feet (1,200 and 1,350 m) constitute the upper edge of the desert. The characteristic species of the bajada are here confined to southerly slopes, and all but half a dozen of them find their uppermost limits. At the upper edge, adjacent to the lower encinal, Arizona rosewood becomes common, although confined to ledges of rock. Alligator juniper, Mexican blue oak, and Arizona whiteoak occur for the first time away from canyons. On the northerly slopes, sotol occurs in abundance, together with the lowest individuals of beargrass, manzanita, amole, and banana yucca.*

The physiognomy of the upper desert slopes is made distinctive from that of the lower desert slopes not only by the entrance of these plants of striking form and the exit of the desert species but also by the abundance of perennial grasses, root perennials, and small shrubs, which combine with the ephemeral plants, or their dead remains, to give a much more complete ground cover than is to be found in any part of the bajadas. There are many scattered species of grama grass along with muhly, feather grass, and tanglehead. There is also a multitude of small shrubs spotted throughout the area.

## Encinal Region

### Lower Encinal [†]

The activity of growth and flowering, so conspicuous on the desert in the season of winter rains, is much reduced in the lower encinal. Extremely few of the ephemerals that often carpet the desert in January are to be found in the encinal region. There is some activity on the part of root perennials during the months of March and April, and flowers may be found on species of mallow, mariposa lily, verbena, penstemon, wild buckwheat, and mustard.

The ground cover of low perennial plants, grasses, succulents, and herbaceous species that has been mentioned as char-

---

*Lloyd observed, "Higher up and forming in places a nearly continuous mat covering many acres [hectares] in extent, is a smaller species of agave, the 'amole' *(A. schottii)*. One who climbs up to any height in these mountains becomes only too well acquainted with this plant, for its leaves are armed with a sharp spine and are placed at just the right height to puncture one's shin and knees in trying to pass over it. It seems to one going up hill as bad as it can well be, but is really a very great deal worse when coming down. This is one plant with which we could very well do without" (Lloyd, 1907b).

†The Spanish word "encinal" signifies a grove or forest of evergreen oaks, being derived from *encina* (evergreen oak). Shreve did not use the terms chaparral or oak woodland in this study but did use them in other publications.

acterizing the upper desert is likewise to be found throughout the lower encinal, but it does not form so close a carpet.

The species that chiefly characterize the lower encinal at its desert edge have already been mentioned. These become more abundant at 5,000 feet (1,500 m) except for blue oak, a tree of very narrow vertical range, rarely occurring above 5,200 feet (1,560 m) and reaching its limit at 5,600 feet (1,680 m) on steep south slopes. Emery oak and pinyon make their appearance at 5,000 feet (1,500 m) along with silk tassel, catclaw mimosa, skunkbush sumac, and redberry buckthorn.

The only characteristic desert species that persist throughout the lower encinal are saguaro—of which Shreve reported a single young individual at 5,100 feet (1,530 m)— staghorn cholla, which reaches 5,500 feet (1,650 m), and fishhook mammillaria, which ascends to 7,000 feet (2,100 m). Two species of prickly pear are found throughout the encinal, growing in thin soil or on rocks and reaching their highest occurrence solely on ridges or upper slopes. One has been found on a sharp rocky ridge at 7,200 feet (2,160 m), which is the highest known occurrence of the prickly pear in the Santa Catalina Mountains. Fishhook mammillaria ranges from the upper desert to nearly 7,000 feet (2,100 m). Hedgehog cactus ranges from about 5,000 to 7,800 feet (1,500 to 2,340 m), which is the highest elevation at which any cactus has been found in these mountains.

## Upper Encinal

Shreve reported that during the ascent from 5,000 to 6,000 feet (1,500 to 1,800 m), the most notable change in the vegetation was the gradual increase in the density of the stand of evergreen trees and shrubs—a change that forms the chief distinction of the upper encinal from the lower encinal. Emory and Arizona oak are still the dominant trees, while pinyon pine and alligator juniper are somewhat less common. Manzanita and silk tassel are the most common of the larger shrubs, and catclaw mimosa of the smaller ones. Sotol, beargrass, and Palmer agave remain abundant—at least on southern slopes, up to 6,000 feet (1,800 m)—and amole remains common up to its upper limit at that elevation. With the increasing abundance of the oaks, however, these semi-desert species as well as the cacti become infrequent and are confined to the ridge summits and the rock crevices.

On steep north slopes, between 5,300 and 6,000 feet (1,590 and 1,800 m) are found many almost pure stands of pinyon and also the lowest individuals of netleaf oak, here a low branched tree of 20 feet (6 m) in height. Chihuahuan pine first appears at about 5,900 feet (1,770 m) on south slopes, being the only one of the trees that is not found at much lower elevations.

The heaviest stands of the upper encinal constitute a relatively dense thicket in which trees are from 18 to 30 feet (5.5 to 9 m) in height and so closely placed that it is very difficult for a mounted man to make his way among them. The upper encinal contains the same species of trees, shrubs, and larger perennials that form the very open lower encinal. There are, however, many root-perennial herbaceous plants in the upper encinal that are not to be found below 5,500 feet (1,650 m) nearly all of which extend upward into the lower portions of the forest region.

The vegetation of the upper encinal is extremely poor in shrubs of the type so common in the upper desert and still frequent in the lower encinal, but vegetation of rocks and exposed ridges is still suggestive of the desert, both in its physiognomy and in its phyletic relationships.

## Forest Region

One of the most striking changes encountered in the vegetational gradient of the Santa Catalina Mountains is from the closed and relatively low encinal to the open forest of Arizona pine, with trees 50 to 60 feet (15 to 18 m) in height. This pine is closely related to ponderosa pine—which Shreve called the western yellow pine—and is the common tree of the forested altitudes of the mountain, extending upward on southerly slopes to the summit of Mount Lemmon.

The lowest stands of sufficient density to be regarded as

forest occur in northerly slopes 5,800 to 6,000 feet (1,740 to 1,800 m) or on southerly slopes at 6,000 to 6,400 feet (1,800 to 1,920 m). The limits depend in each particular locality upon the steepness of the slope and its soil characteristics, particularly with respect to moisture supply.

White fir occupies the northern slope of the highest summits and ridges of the range from 7,500 feet (2,250 m) upward, but there are no elevations in the Santa Catalina Mountains sufficiently great to bring the fir forest onto the south slopes.

In the forest region, winter is a season of almost absolute rest, save for the photosynthetic activity that is doubtless carried on by the conifers, and possibly by the evergreen oaks and shrubs. Deciduous trees and shrubs are leafless from early or mid October until April or May. Only a few herbaceous perennials are active during this period, for instance the evergreen species of *Pyrola* and such early vernal plants as elkweed. Perennial herbaceous activity during the arid foresummer is largely dependent on the amount of winter precipitation and date of its termination. In the lower portion of the pine forest, almost all activity is in abeyance until the first rains of the humid midsummer, while in the upper pine forest it is always possible to find the majority of the common herbaceous plants in activity in May and June. There is notable scarcity of annual plants above 6,000 feet (1,800 m).

## Pine Forest

In the lowest part of the pine forest many of the dominant encinal forms are still to be found, but evergreen oaks become more scattered in occurrence. Emory oak and pinyon are scarcely overlapped; manzanita and silk tassel are likewise of infrequent occurrence in stands of forest. The oaks characteristic of the closed forest are netleaf oak and the silverleaf oak. The former is commonly a low-branching shrub that often forms thickets on the steep slopes of the highest peaks where it extends upward to about 8,600 feet (2,580 m). The latter oak is a shrub near its lower and upper limits at 6,000 and 8,000 feet (1,800 and 2,400 m) but attains a height of 40 feet (12 m) and a girth of 3 to 4 feet (1 to 1.2 m) between 6,500 and 7,500 feet (1,950 and 2,250 m). Alligator juniper is of occasional occurrence in the forest up to 7,900 feet (2,370 m), and Arizona madrone, at first infrequent, becomes common at 7,000 to 7,500 feet (2,100 to 2,250 m) and reaches its upper limit at 7,800 to 8,000 feet (2,340 to 2,400 m).

According to Shreve the composition of the forest itself is extremely simple from its lower limit around 6,000 to 7,500 feet (1,800 to 2,250 m). Above that elevation the composition is equally simple on southerly slopes up to the summit of Mount Lemmon. Chihuahua pine reaches its limit at about 6,700 feet (2,010 m) and forms a very inconsiderable portion of the forest throughout the uppermost 500 feet (150 m) of its vertical range. Douglas fir begins to occur on steep northerly slopes at 6,100 feet (1,830 m), and Mexican white pine at 6,800 to 7,000 feet (2,040 to 2,100 m), but neither begins to affect general composition of the forest until higher elevations are reached.

At 6,000 feet (1,800 m) the streamways and floodplains are characterized by several deciduous trees in addition to the pines. The sycamore is near its upper limit at this elevation and the native Arizona black walnut, southwestern chokecherry, and mountain maple are of frequent occurrence.

Throughout the pine forest there are trees, shrubs, and herbaceous plants that may be found in the encinal, but only in the lowest edge of the pine forest may plants be found that suggest the genera or vegetation types characteristic of the desert. Hedgehog cactus, a yucca, and an agave are the sole representatives of the succulent and semi-succulent forms of the lower elevations.

The pine forest does, however, have vegetational features that suggest the effects of a climate not far removed in character from that of the desert. The openness of the lowest stands of Arizona pine, the high mortality among the seedlings of the pine, the character of the foliage of these plants, all point to the existence of a precarious soil-moisture supply and to atmospheric conditions conducive to active transpiration. In the fir forest none of these features is observable, and the vegetation as a whole presents a much more mesic aspect.

The pine forest gives the impression of a much richer flora of herbaceous plants than is found in any other habitat of the mountain. This impression is due to the large number of species that enter into the vegetation as very common components. The total flora, however, is not so great as might be supposed on first examination.

## Fir Forest

Between 7,000 and 7,400 feet (2,100 and 2,220 m) there is a rapid change in the character of the forest stands on northerly slopes, owing to the increasing occurrence of Douglas fir and Mexican white pine, the lower limits of which have already been mentioned, and to the appearance of white fir (see Fig. 4.1). These three species occur in mixed stands together with Arizona pine on northerly slopes up to about 7,500 feet (2,250 m). Above this elevation the latter becomes a very infrequent tree on slopes facing directly north, although it still occurs in a mixture with Douglas fir and true fir at 9,000 feet (2,700 m) on eastern and western exposures. In the fir forest the last relics of the encinal have disappeared. Silverleaf oak, netleaf oak, and alligator juniper are nowhere to be found, although they may grow very nearby on opposed slopes. Arizona madrone is also absent from the fir forest. The deciduous Utah white oak and the widely distributed quaking aspen are the commonest of the subordinate trees, the latter often becoming dominant over areas of an acre (.4 ha) or more in extent, where it ultimately gives way to conifers.

The floor of the fir forest is much more heavily and continuously shaded than that of the densest stands of pine, a circumstance which is of great importance in determining the nature of the forest reproduction and also in conditioning the character of the shrubby and herbaceous vegetation. The dense shade, the heavy litter, and the high humus content of the soil tend to preserve its moisture throughout the arid foresummer, so that seedling trees and other plants of these situations are very far removed from the desiccating influences that are operative in the open pine forest. The heaviest stands of white fir and Douglas fir, like most heavy coniferous forests, are relatively poor in both shrubs and herbaceous plants.

The banks of constant and intermittent streams and the narrow floodplains of the fir forest region form a series of habitats with closely similar physical conditions and with nearly identical vegetation. In them are to be found a greater abundance and variety of trees and shrubs than occur in topographically analogous habitats at lower elevations. Common woody plants include those that do not occur in other situations such as alder, maples, willows, dogwood, sambucus, snowberry, dewberry and orange gooseberry.

## Santa Catalinas Compared With Other Mountains

To follow up his studies on the vegetation of the Santa Catalina Mountains, Shreve (1919) made a comparison of the vegetational features of the Santa Catalina Mountains and the Pinaleno Mountains (Mount Graham) near Safford, Arizona. These two mountains had a great deal of similarity, especially as they were both relatively in the same stage of physiographic development, both built chiefly of gneiss, and both in a WNW-ESE position. It was possible, therefore, to compare the vegetation at the same elevations of the two mountains with a minimum of complicating features.

Part of Shreve's study was to determine the influence of the greater altitude of the Pinaleno mountains on the character of vegetation and flora, the influence of slope exposure, and the influence of vertical distribution by the elevation of the surrounding country. He noted that possibly the most important physical features of the Pinaleno Mountains were determined by the fact that on its northeastern side its drainages reached the Gila River at elevations of about 2,600 feet (780 m), and on the southwest side its stream fell to the Bonita Valley at only 5,000 feet (1,500 m). Differences in vegetation were related to the base level on these two sides.

The larger features of the vertical distribution of vegetation in the Pinaleno Mountains are similar to those of the Santa Catalina Mountains. Dissimilarity between the two mountains is due to the greater altitude of the Pinaleno Mountains by 1,400 feet (420 m) and to the fact that the Pinaleno canyons are more sharply cut and better watered. The higher elevation results in extended areas of forest of a type only sparingly represented at the highest altitudes in the Santa Catalina Mountains. The rugged topography of the Pinaleno Mountains and the numerous constant streams resulted in a sharper contrast between the vegetation of canyon and slopes of lower and middle altitude.

The gently rolling summits of the Pinaleno range lie chiefly above 9,500 feet (2,850 m) and are clothed with a fir-spruce forest to a much greater extent than the analogous portion of the Santa Catalina Mountains. Numerous canyons that are very precipitous have eaten into the edge of the summits of the Pinaleno Mountains. This has limited the extent of pine forest and has presented conditions favorable for the high occurrence of the trees and shrubs that characterize the encinal. The well-watered canyons and steep slopes have resulted in a pronounced interdigitation of the highland-lowland vegetation so that plants of streamways are carried nearly 1,000 feet (300 m) lower than they are in the Santa Catalina Mountains, and the encinal is carried about 1,220 feet (360 m) higher than in the Santa Catalina Mountains. The lofty and sharply dissected alluvial aprons on the northeast side of the mountain extend down to an elevation of 2,800 feet (850 m) at the Gila River and plains exist on the southwest side at 5,500 feet (1,650 m). As a result, there are marked dissimilarities in the vegetation on the two faces of the mountain. Another difference between the Santa Catalina Mountains and the Pinaleno Mountains is that many plants of Sonora and Sinaloa, characteristic of the lower elevations of the Santa Catalina Mountains, are not found in the Pinaleno Mountains. At higher elevations many plants common to the Rocky Mountains of northern New Mexico are found in the Pinaleno Mountains, but not in the Santa Catalina Mountains. Shreve concluded that, owing to the similarity of climate and other environmental conditions in the two mountains, the absence of these species from the Santa Catalina Mountains would appear to be due to causes other than those of physical environment (Carnegie Yearbook 14).

In a study comparing the Santa Catalina, Pinaleno, Santa Rita, Huachuca, and Whetstone Mountains, particular attention was given to detecting the plants that are common components of the vegetation in the Pinaleno and Huachuca Mountains but are not known to occur in the Santa Catalina Mountains. The Huachuca Mountains are well known to have a rich flora, comprising a score or more of plants that have not been seen elsewhere in the United States, although some of them are common to the Sierra Madre Occidental of Mexico. A few of these forms are conspicuous elements in the vegetation of the Huachucas. The Huachucas are also the location of a number of species of high-mountain plants common to northern New Mexico and Colorado that do not occur in the Santa Catalina Mountains (*Carnegie Yearbook 16*).

Shreve believed that distributional movements of plants among the isolated mountains of the Southwest have been controlled by the mobility of the plants themselves and by the physical conditions which their invasions have encountered. A few cases discovered in the Santa Catalina Mountains are believed to represent early stages of invasion by plants that are widespread in adjacent mountain areas. The principal evidence is the fact that they are now occupying only an extremely small part of the terrain favorable to them and that they show no tendency to localized occurrence in the adjacent mountains. Among these are corkbark fir, cottonwood, grouse whortleberry, bitter cherry, and Virginia creeper.

The influence of altitude per se (that is, differences in barometric pressure) has yielded no conclusive demonstration of effect on vertical distribution of plants on desert mountains. It is the altitudinal differences of insolation, temperature, rainfall, humidity, evaporation, and a score of related conditions that make up what has been called the "factor of altitude" (Shreve, 1922).

On an ideal mountain, having the form of a cone, with uniform gradient from base to top, without irregularities of surface, and with identical soil throughout, it would be possible to determine very clearly the relation between the vertical ranges of the biota and the equally ideal gradients of conditions. Under the natural conditions, however, it is necessary to make many interpretations of observed phenomena because of

variation in the normal gradient conditions and local departures from the normal.

Shreve stated the most common way in which the topographic irregularities of a mountain may cause departure from
the ideal vertical distribution is simply to raise or lower the
limits. Thus, in all desert mountain ranges there is a marked
difference in the ridges or peaks and of the bottoms of valleys
at the same elevation. There is also a striking difference between the vegetation of north and south slopes at all elevations.

# Desert Climate

## *Precipitation*

DESERT CLIMATE HAD, for some time before the establishment of the Desert Laboratory, been a topic of considerable importance, because it related to the extreme lows of moisture conditions. There were varying opinions as to what constituted the upper limits of desert moisture, but these, fortunately, had not interfered with the collection of a large amount of data on "arid" climates—varying all the way from extreme dryness to subhumid conditions.

Ellsworth Huntington, a frequent visitor at the Desert Laboratory, described the climate of Arizona, New Mexico, and northern Sonora as "subtropical continental" and of the monsoon variety. It resembles that of the provinces of the Punjab, Rajputana, and Sind in northern India. The region extends from about north latitude 28° in Mexico to 37° in northern New Mexico and Arizona. Its subtropical position brings most of it within the great world zone where high pressure and consequent aridity normally prevail. The main movement of the air is downward and outward. The northeasterly winds of the trade-wind zone and the southwesterly winds of the prevailing westerlies zone find their origin in this region.

101

The cyclonic storms of the westerlies in winter and the descending air of the subtropical "horse latitudes" in spring and autumn give rise, respectively, to the rain and aridity that would be expected. In summer, however, because of the great size of the continent of North America, the trade winds that would be expected do not appear; their place is taken by relatively moist winds which blow, in general, from the south, and may be called monsoons for lack of any more appropriate name.

This southerly monsoon gradually becomes well established after the dry foresummer by the strong indraft toward the heated continent, and thundershowers finally begin upon the mountains. Far to the south, in Mexico, the first showers may come in May or even April. In southern Arizona, they usually begin toward the end of June or early in July, while farther north, they do not come until mid-July. In exceptionally warm years, however, they may begin unusually early because of the more rapid heating of the continent. Everywhere they are accompanied by vivid lightning and torrential rainfall.

Ellsworth Huntington pointed out that the rainfall is generally less than 12 inches (30 cm) and may be as low as 3 inches (7.5 cm). Regions having a rainfall of only 10 or 12 inches (25 or 30 cm) may be bare and treeless—as can be seen in Utah or Nevada, or in Syria and Persia. Their arboreal vegetation, away from the watercourses, is almost entirely restricted to insignificant grayish-green forms like sagebrush. In the southern part of Arizona, on the contrary, bushes and trees are found almost everywhere, and the aspect of the desert is distinctly arboreal and verdant.

Huntington believed that the peculiarly verdant arboreal character of the desert of southern Arizona and Sonora is due primarily to a double rainy period. In the majority of deserts, rain falls only during a single season—often the winter—when temperatures are unfavorable to growth. In Arizona, where both winter and summer rainfall is common, the desert supports good growth of herbaceous annuals in summer. Trees require a prolonged season of growth, and the brief moist season in most deserts does not fill the ground with sufficient moisture for trees to mature their various organs and produce seed. In Arizona the winter rains start the growth of trees and supply sufficient moisture for the plants to subsist until the arrival of summer rains. These then lengthen the growing season to equal that in many regions which are much better watered. Moisture is still scarce for a long interval during the rainless foresummer, and the ground is too dry for ordinary trees. Nevertheless, many desert species have become adapted to the double rainy season. Hence, although Arizona is a genuine desert from an agricultural point of view, the scenery of the southern part by no means suggests this character. The country is far more verdant than many regions where agricultural possibilities are much greater.

Because of its relative scarcity and uncertain distribution, rainfall is undoubtedly the most important climatic factor to be considered. With this in mind, a record of rainfalls was kept at the Desert Laboratory from its establishment in 1903. The following year a rain gauge was placed in the foothills of the Santa Catalina Mountains, some 10 miles (16 km) distant from the Laboratory. Numerous other rainfall stations were established from time to time and were maintained for longer or shorter periods in connection with experimental projects.

### Early Investigations

In 1925 an extensive investigation of the rainfall in the Sonoran Desert was begun. Two principal lines of rain gauges were established along diverging routes from Tucson, and a series of gauges were concentrated on the grounds of the Laboratory. The series of gauges known as the Libertad Line extended in a southwesterly direction from Tucson to Cirio Point, formerly designated as Kino Point, 9 miles (14 km) south of Port Libertad, Sonora, on the Gulf of California. Each gauge was located at a point that possessed some particular botanical or topographical feature. The Camino Line of gauges followed the route that leaves the Libertad Line 24 miles (38 km) west of Tucson. This route continued on to Ajo, then south to the international boundary, and then in a north and northwest direction along the old Camino del Diablo emigrant trail. The farthest rain gauge was 10 miles (16 km) south of Wellton, Arizona, and approximately 40 miles (64 km) southeast of

Yuma, Arizona. The most distant gauges on the Libertad and Camino lines were each approximately 245 miles (392 km) by road and 180 miles (288 km) by air line from Tucson.

In order to obtain these rainfall records, it was necessary to develop a rain gauge that could be read at infrequent intervals. The rain gauges were read twice a year, usually during the months of April and October, unless unusual conditions and other uncontrollable circumstances made this impossible. The gauges used were developed by Sykes (1931). The instrument designed consisted of a truncated conical vessel about 9.5 inches (23.7 cm) in height, and the same diameter at the base made of sheet copper folded with well-soldered seams, surmounted by a vertical sided but conical bottomed measuring bowl 3 inches (7.5 cm) in diameter, fastened to the upper end of a short connecting tube that passes through a removable screw cap, well down into the bottom of the container. A short spout is fitted at the angle of the frustrum-shaped body. Copper was used in construction as it is more durable than sheet iron or tin plate. It was found in practice about 0.25 pint (100 cc) of a light mineral oil fully protected the surface of the contained water against evaporation for a period of a year or more even under trying conditions of desert sunshine. This instrument could contain as many as 11 pints (4500 cc), corresponding to 40 inches (100 cm) of rainfall, which was found to be more than ample for even mountain-top reading upon the basis of the usual six-month period. It was found expedient to paint the outside of the instrument and also the inside of the bowl a dull drab color and to mix paint with fine sand in order to make the apparatus less conspicuous to chance passersby.

*Since this principle of camouflaging has been adopted, very few of the instruments have been interfered with except in one case where a nearly completed six month record was lost through the misplaced zeal of two ardent prohibition agents, who, finding it an unknown object made of copper, assumed that it must be a new form of still. [Sykes, 1931]*

Dwight Mallery (1936a, b), who made most of the rain-gauge observations, found the most satisfactory vessel to use in measuring the rainfall was a 1-liter graduate brass cylinder (Fig. 5.1). Also, after the gauge was emptied, it was rinsed

Fig. 5.1 Rain gauge used in Desert Laboratory investigations.

thoroughly with kerosene to remove any accumulation of surplus oil, insects, lizards, and so forth. The effectiveness of the oil seal in preventing evaporation of the rainwater between readings was tested by placing the gauge side by side with a Standard Weather Bureau rain gauge for one year. Very close agreement was recorded for both gauges (Mallery, 1936a).

Much care was taken to place the gauges in spots somewhat removed from roadways and unfrequented by souvenir hunters. Buried up to the necks in soil or rocks, with only the 3-inch (7.5 cm) funnel protruding, the gauges were not easily detected. Only a few were lost during the entire epoch of the long-period gauges, but in one instance a Papago squaw was seen utilizing the lower portion of a gauge as a teakettle! How could she have known its importance in a scientific investigation?

Wild animals were curious about such unusual objects but due to the conical shape of the gauges rarely succeeded in exhuming them. It is claimed by naturalists that the coyote is the most intelligent animal on the North American Continent and possibly one of the most fastidious in remote places. However, the coyote was a factor that must be kept in mind when evaluating the accuracy of the readings. This was especially important when there were signs that some practical coyote had utilized the gauge as a latrine! Only the experienced technician could judge to what extent such use might vitiate the long-period rainfall totals.

Not only did the invention of the long-period gauge contribute to our knowledge of desert rainfall, but its use helped to keep the Desert Laboratory staff and other scientists in touch with the desert and conditioned for intensive and extensive ecological expeditions throughout the Sonoran Desert. These investigations evolved as the principal project of the Desert Laboratory until discontinued owing to the economic depression of the thirties.

Table 5.1 and Figures 5.2 and 5.3 adapted from Mallery (1936a, b), summarize the rainfall readings of this period. Although the readings were limited to a few years, they are believed to be indicative of relative precipitation. In regard to the greater amount of rainfall at most stations in the summer, Mallery pointed out that the amount of rain in summer and winter tends to become equal to the west, until winter rains in the extreme southwestern corner of Arizona exceed summer rains in quantity.

## Rainfall in the Sonoran Desert

Turnage and Mallery (1941) carried out a very comprehensive and detailed analysis of precipitation in the Sonoran Desert and adjacent territory.

They used the rainfall stations that had been maintained by the Desert Laboratory in southwestern Arizona and northwestern Sonora. These records were for a period of less than ten years and were considered to be not as reliable as desired, but they did have the advantage of being located at strategic points in the Sonoran Desert. In addition to the Desert Laboratory stations the authors utilized the records from the U.S. Weather Bureau stations varying in length from twenty to forty years, and some for even longer periods. In a very few

Table 5.1. Seasonal and Annual Rainfall as Recorded by Long Period Rain Gauges in Southwest Arizona and Northwest Sonora (from Mallery, 1936a, b)

| Station* | Length of Record (years) | Seasonal Range | | | | Seasonal Average | | |
|---|---|---|---|---|---|---|---|---|
| | | Summer | | Winter | | Summer (in.) | Winter (in.) | Ratio (Summer=100%) |
| | | Max. (in.) | Min. (in.) | Max. (in.) | Min. (in.) | | | |
| 1. Cirio Point | 10 | 11.17 | 0.43 | 2.90 | 0.27 | 2.41 | 1.57 | 65.1 |
| 2. Puerto Libertad | 9 | 8.93 | 0.62 | 4.00 | 0.39 | 2.31 | 1.66 | 72.0 |
| 3. 19 Mi. Pass | 8 | 14.48 | 4.14 | 4.73 | 0.48 | 8.20 | 2.48 | 30.2 |
| 4. 50 Mi. Pass | 9 | 19.64 | 4.39 | 3.62 | 0.62 | 10.06 | 1.76 | 16.6 |
| 5. Los Temporales | 3 | 8.39 | 4.24 | 1.20 | 0.67 | 5.94 | 0.93 | 15.6 |
| 6. Oquitoa | 2 | 6.46 | 4.91 | 3.80 | 0.47 | 5.68 | 2.13 | 37.5 |
| 7. Red Rock Crossing | 3 | 14.46 | 6.42 | 4.22 | 0.85 | 9.65 | 2.94 | 30.5 |
| 8. Los Molinos | 2 | 8.01 | 7.58 | 5.77 | 3.80 | 7.75 | 4.78 | 61.3 |
| 9. Baboquivari | 4 | 16.21 | 8.66 | 9.05 | 3.53 | 13.30 | 5.96 | 44.8 |
| 10. Sierrita Mts. | 7 | 23.75 | 3.75 | 7.32 | 1.83 | 10.94 | 4.94 | 45.2 |
| 11. Avra Valley | 6 | 11.61 | 2.68 | 5.95 | 1.85 | 6.65 | 3.46 | 52.0 |
| 12. Soldier Camp | 7 | 29.82 | 9.82 | 24.64 | 4.27 | 18.51 | 16.22 | 87.3 |
| 13. Pima Canyon | 5 | 14.28 | 3.36 | 9.00 | 1.25 | 8.47 | 4.75 | 56.1 |
| 14. Desert Lab. grounds: | | | | | | | | |
| A. Near N. gate | 7 | 14.73 | 4.04 | 7.33 | 1.46 | 7.95 | 3.99 | 50.2 |
| B. Midway between N. & S. limits | 7 | 9.28 | 4.46 | 7.93 | 1.64 | 6.50 | 4.87 | 74.9 |
| C. N.W. corner | 5 | 8.12 | 3.87 | 8.20 | 1.85 | 5.77 | 5.20 | 90.2 |
| D. S.W. corner | 5 | 8.20 | 3.62 | 7.37 | 1.81 | 5.85 | 4.70 | 80.4 |
| E. S. of Tumamoc Hill | | 9.50 | 3.72 | 7.40 | 1.77 | 6.24 | 4.77 | 75.8 |
| F. Summit of Tumamoc Hill | 9 | 13.66 | 3.23 | 7.93 | 1.56 | 7.57 | 4.21 | 55.6 |
| 15. Sells | 3 | 12.07 | 5.27 | 4.90 | 2.93 | 7.71 | 3.91 | 50.7 |
| 16. 23 Mile Hill | 6 | 10.53 | 3.88 | 5.35 | 2.15 | 7.13 | 4.05 | 56.4 |
| 17. Growler Pass | 3 | 5.86 | 2.80 | 6.78 | 1.51 | 4.37 | 4.39 | 100.4 |
| 18. Agua Dulce | 9 | 6.79 | 2.15 | 6.40 | 1.29 | 3.72 | 3.66 | 98.7 |
| 19. Pinacate Plateau | 7 | 7.00 | 0.89 | 6.25 | 0.13 | 3.35 | 2.76 | 82.4 |
| 20. Tule Tank | 8 | 6.43 | 0.00 | 4.22 | 0.00 | 2.38 | 1.94 | 81.5 |
| 21. Tinajas Altas | 8 | 5.08 | 0.31 | 4.50 | 1.25 | 2.45 | 2.83 | 115.5 |
| 22. Lechuguilla Desert | 3 | 2.41 | 0.36 | 4.22 | 0.95 | 1.26 | 3.06 | 242.9 |

* Most station locations are shown in Figure 5.2.

107

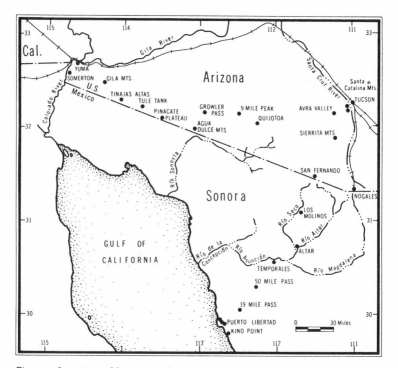

Fig. 5.2. Location of long period rain gauges (see Table 5.1). Adapted from the *Geographical Review*, Vol. 21, 1931, with the permission of the American Geographical Society (Sykes, 1931).

cases did the ten-year average depart as much as 10 percent from the twenty- to forty-year average, a fact which they believed justified the use of all available records to draw their conclusions.*

In this study, the records were divided into seasons: sum-

---

*There is no question as to the validity of these records, but Harold Fritts (University of Arizona, personal communication), after researching the course of annual precipitation variations on the basis of tree ring studies and other sources of information, has stated that the period covered in this study represented above average precipitation conditions. He does not dispute the seasonal relationships but cautions that the overall precipitation figures are higher than the average.

Fig. 5.3. Rainfall record from the Sonoran Desert (after Mallery, 1936a,b). Distribution and approximate location of collecting stations is shown in the inset, and names of numbered stations are given in Table 5.1. Stations 14 and 15 were omitted from the graph because their location in the Santa Catalina Mountains was outside true desert conditions.

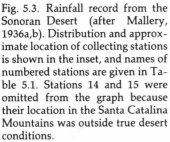

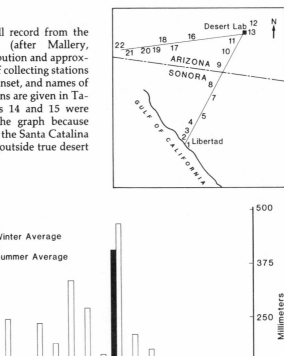

mer, including May through October; and winter, including November through April.

Winter rainstorms in Arizona usually began in November and continued into March. These rains were apparently continuous with storms of the western United States. They were associated with the passage of low-pressure areas and fronts across the continent. These prolonged storms covered a large area, occurring over the entire state at the same time; however, a rain of more than thirty hours duration was rare. Precipitation was well distributed with the respective time of day, although the nighttime hours produced somewhat more rain. Rainfall intensity (in Tucson) was low, more than 0.15 inch (0.38) cm) in a fifteen-minute interval being uncommon. More than half of the total winter precipitation fell at a rate below 0.05 inch (0.12 cm) per fifteen minutes.

The bulk of the summer rainfall came in July, August, and September in brief showers usually scattered over the entire state for several days, followed by several days of dry weather. Showers occurred locally; even in the wettest weather, sizable dry spots separated the wetted areas. Practically all the showers took place between noon and midnight, the amount of precipitation increasing toward sunset. Heavy rains often developed in midafternoon from a sky that was cloudless at dawn. The showers were usually of short duration but were very intense; an intensity of 0.75 inch (1.9 cm) in fifteen minutes had been measured at Tucson. In spite of these high intensities and the spectacular thunderstorms that were quite common, about half the average summer rain fell at less than 0.10 inch (0.25 cm) in fifteen minutes. The highest rainstorm intensity was usually near the beginning of the storm, and many rains concluded with several hours of relatively gentle precipitation.

Records used by Turnage and Mallery show practically all of Sonora, southwestern Arizona, the southeastern corner of Arizona, the California portion of the Sonoran Desert, and the northeastern portion of Arizona has from 0 to 5 inches (12.5 cm) of winter rain.

The Arizona boundary of the Sonoran Desert is in an area receiving 5 to 10 inches (12.5 to 25 cm) of winter rain. But the boundary in Sonora lies in an area receiving 0 to 5 inches (0 to 12.5 cm).

The area having 0 to 5 inches (0 to 12.5 cm) of summer rain included the California portion of the Sonoran Desert, southwestern Arizona, a narrow belt along the Colorado River north of the 35th parallel, a small strip along the Little Colorado River north of Winslow, and the coastal region of Sonora, northwest of Libertad. In northern Arizona, the rain belt along the Sonoran Desert boundary, southeastern Arizona, and the central part of Sonora from the Arizona boundary southward beyond Guaymas had from 5 to 10 inches (12.5 to 25 cm) of summer rainfall. Several small isolated areas in western Arizona also had from 5 to 10 inches (12.5 to 25 cm), including Ajo, Kofa, the Harquahala and Silverbell mountains. Considering spatial variation in rainfall over the entire area, a compilation was made of twenty-eight stations for twenty years, supplemented by eighteen stations in southwestern Arizona and northwestern Sonora for the previous five winter and summer seasons. A brief summary was made of each winter and summer season by year. In this summary any reading between 80 and 100 percent was considered normal—not one of the winter seasons could be characterized as such. The seasons that most nearly approached normal either were very spotted or had only half the area dry. Seven seasons were classified as wet and eight as dry. Therefore, three-fourths of the winter seasons were either wet or dry and the remaining one-fourth were much better described as irregular than as normal. Summer seasons were much more difficult to classify than were the winter seasons with respect to spatial variation for the entire area. A large number of the summer seasons were highly spotted. Five seasons were definitely wet and two seasons were half wet and half dry. Only half the summer seasons could be considered either wet or dry, and no season showed values uniformly near the mean.

A study was made of an area near Tucson including four stations: The University of Arizona; 3 miles (5 km) east of the Desert Laboratory; the then Airport station, 4.3 miles (6.9 km) southeast of the University of Arizona; and the Magnetic Observatory that was 7.5 miles (12 km) northeast of the University. The study showed that the seasonal summer rainfall was much more erratic than the winter rainfall with respect to spatial variation, where the area is somewhat larger than 1 square mile (259 ha). However, this is not always the case.

That the amount of rainfall varies directly with elevation has become somewhat of a generalization. However, an analysis in the Southwest indicated the limitations of this apparent correlation. Several conditions may mask or accentuate the influence of elevation upon rainfall. These conditions were identified as geographical location of the station, the direction of the rain-bearing wind, the relief of the surrounding area, and the meteorological unity of the storms.

Turnage and Mallery observed that the amounts of seasonal rainfall in Arizona and the Sonoran Desert are often influenced by the topography of the land immediately surrounding the station. Perhaps the best-known illustration is the rain-shadow effect of large mountains on the lower land to leeward, but certain stations showed a positive departure from the expected precipitation for their location and altitude and were believed to be influenced by the relief of the land surrounding the station. In the winter season the Roosevelt station had about 4 inches (10 cm) more rainfall than a station at its elevation would normally receive. They thought this excess was probably due to the presence of the Mazatzal Mountains, which reach an altitude of 7,300 feet (2,190 m) a few miles west of the station. They are relatively high but of too small mass materially to deplete the clouds of moisture. The westerly rain-bearing winds are deflected upward by these mountains, but they carry over sufficient moisture to give Roosevelt, on the leeward slope, an abnormally high rainfall. At the same elevation but 10 miles (16 km) east, only the rain predicted for this region was received, indicating that the mountain influence did not extend that far. Much the same story explains the difference in precipitation between Miami and Globe. Both are to leeward of the 9,000-foot (2,700-m) Pinal Mountains, but Miami, with average precipitation of about 19 inches (48 cm) is at the base of the mountain mass and Globe with an average of about 16 inches (40 cm) is more than 5 miles (8 km) beyond.

### Rainfall in Mexico

The scientists at the Desert Laboratory did not make any rainfall measurements outside of the Sonoran Desert, but Shreve (1944), using secondary information, brought together information on the rainfall of northern Mexico. The major source of precipitation data was the publication of the Mexican Meteorological Service, which had been in continuous operation for over fifty years. Records for state capitals and other large cities were found to be excellent. There were also records varying in length and completeness kept by voluntary observers, ranches, mines, and lumber camps.

The geographical structure of northern Mexico is simple in its larger features yet very intricate in its details. The complex interplay of climatic conditions controlling vegetation gives little basis for predicting the vegetation in one area as based on conditions in another.

Broad plains rise from the coasts on the east and west. The Sierra Madre ranges stand parallel to each of the coasts on the eastern and western sides. Between these ranges lies the central plateau that rises gradually from the Rio Grande. Although much of the plateau is flat or rolling, it also supports many isolated hills and mountains.

Throughout northern Mexico, except the extreme northwest, the principal rainfall occurs during the months May to September. The rain-bearing storms have their origin in the Caribbean area with the advent of summer. The gradual march of the rainy season across the Mexican plateau determines the monthly distribution of precipitation. May is marked by heavy rain in Chiapas, Tabasco, and Veracruz, as well as in northern Tamaulipas. The rest of the country is relatively dry as far south as Mazatlán, and there is no May rain in the lowlands of Sonora and Baja California. In July heavy rains are more widespread and light rains extend into Sonora and southern Baja California. August conditions are very similar to those of July, except that many localities in the northeast show the midsummer depression characteristic of that region and no rain falls in southern Baja California. September is the month of maximum rainfall for most of the Gulf of Mexico and in and near the eastern Sierra Madre. Rain then begins to wane in the western Sierra Madre as well as in northern Chihuahua and the lowlands of Sonora. The conditions in October are an almost exact return to those of May.

A cross section of northern Mexico exhibits great differences in mean annual rainfall as well as in its monthly distribution. The altitudinal ranges of the types of vegetation found in northern Mexico were largely associated with the local amounts and seasonal distribution of rainfall. As there was no universal relation between altitude and rainfall, there was consequently none between altitude and vegetation.

An important aspect of precipitation in arid and semi-arid regions is the duration and frequency of rainless periods. These were summarized by Shreve for the northern part of Mexico and showed the longest number of consecutive rainless months and the number of periods of three months or more without precipitation. Monterrey, Saltillo, Chihuahua City, and San Luis Potosí all lie just outside the desert and appear to be rarely visited by more than a single month of drought. The stations in the Chihuahuan Desert have registered from three to seven consecutive months without rains. On the west coast in the Sonoran Desert, dry periods of four to seven months had been observed at all stations. The longest period, eight months, was recorded at Mazatlán in the thorn forest. Available records for Baja California indicated rainless periods of seven to eleven months. On the basis of the number of times during a twenty-year record that rainless periods of three months have occurred, the comparative aridity of the Chihuahuan and Sonoran deserts were in the range of two to four periods for the former and three to thirteen for the latter. In Baja California rainless periods of three months were the rule rather than the exception.

## Storms and Droughts

Shreve observed that in all arid regions, exceptionally heavy rains occurred at long intervals. These were more common at low elevations on or near the coasts than they were on the central plateau. Some of these heavy rains are rather spectacular in relation to the average precipitation. At Santa Rosalía, Baja California, for example, the mean rainfall for September was 1.85 inches (46 mm). In September 1939, a single storm of about two hours yielded 7.44 inches (186 mm), which exceeded the annual mean of 4.84 inches (121 mm).*

The irregularity of the rainfall in southern Arizona, and in all desert regions, greatly increases adverse conditions for plants. There appears to be little relation between total annual

---

*Fritts' studies showed that rains during September-October are mostly spent hurricanes or tropical storms or cutoff lows.

rainfall and the amount of moisture in the soil. For example, if a rainfall of 1.20 inches (3 cm) fell in a single shower, or in four rains of 0.30 inch (0.75 cm) or in twelve showers of 0.10 inch (0.25 cm), the effect on the moisture in the soil and of vegetation would be different in each case. In the first, the penetration of water into the soil is not much greater than in a shower of one-third the amount, the remainder running off to the streamways. In the last case there would be only a very superficial wetting of the soil, unless it happened that the showers of 0.10 inch (0.25 cm) followed in rapid succession or fell on cool, cloudy days.

Based on daily entries at the Desert Laboratory, Shreve made a rough separation of the rains into the heavy ones that were accompanied by active runoff, the light ones insufficient to raise the soil moisture, and the intermediate, or effective, rains in which there was little or no runoff and a good penetration of water into the soil.

Shreve pointed out that under desert conditions runoff is greater and more important in its consequences than in moist regions. Heavy rains of brief duration place water on the soil more rapidly than it can penetrate. In all soils of fine texture the dryness of the surface causes puddling as soon as the rain begins, and the penetration of water may advance only a few millimeters by the time the rain is over. Active evaporation quickly removes moisture from the superficial layers of the soil.

An important feature of rainfall conditions is the frequency and duration of the periods in which precipitation is absent or not great enough to affect the moisture of the soil. Shreve considered that a drought period defined by duration of rainlessness was a matter of arbitrary selection. In the warm months a period of two or three weeks without rain will kill the herbaceous ephemeral plants and will check the activity of the root perennials and small shrubs. On the other hand, a period of two or three months without precipitation will be of little consequence to cacti and such deep-rooted trees as the mesquite and catclaw.

Shreve (1934b) selected a length of thirty days as constituting a drought period of mean duration with respect to plant activity. During the twenty-nine years prior to 1933, there were from two to five drought periods annually. The longest period was 149 days, extending from February 16 to July 13, 1928. There were eight periods of more than 100 days, and twenty-four periods of 70 days or more. The distribution of drought periods emphasizes the irregular incidence of the winter rains as compared with those of the summer.

The unusually dry winter of 1924–25 was reported with a precipitation of 0.97 inch (2.42 cm). This was the third year in succession with little or no stand of winter ephemerals. During this very dry period, creosote bush was partially or fully defoliated, the paloverdes had no leaves, the catclaw was leafless, cacti were very thin, there were no flowers on palo-verde, creosote bush or catclaw, and the flowers on the saguaro extended down the sides for 6 feet (2.0 m). The flowers of ocotillo and mesquite were more abundant than usual. Also ironwood had a heavy crop. Flowers of the night-blooming cereus, which almost invariably blooms at the same time each year appeared six weeks earlier than usual (*Carnegie Yearbook 28*).

## Temperature

It didn't take long for the scientists from humid regions to learn that temperature relationships in the desert and the associated desert mountains are very much different from those found in humid areas. The arid regions of the earth exhibit great daily ranges of temperature from a rapid heating of the soil by day and a corresponding rapid cooling through radiation at night. The chief conditions that permit these changes are the absence of a heavy plant cover, the predominance of sand and stones, the usual low air humidity, and the absence of a nocturnal cloud blanket. Rapid nocturnal radiation in the desert is responsible for the phenomenon known as inversion of temperature, which is at once a cause and effect of cold air drainage.

Shreve's studies (1912) on the normal temperature gradient of the Santa Catalina Mountains determined the magnitude of air movement capable of causing variation in the altitudinal limits of species and their seasonal activities in relation to the topographic site.

A series of biweekly readings of the minimum temperatures in the Santa Catalina Mountains between April and October brought out interesting relationships in comparison with
the daily readings of minimum temperatures at the Desert
Laboratory; at the foot of Tumamoc Hill; at the University of
Arizona Experiment Station across the valley from Tumamoc
Hill; and mountain stations at 4,000, 6,000, and 7,000 feet
(1,200, 1,800, and 2,100 m). Another thermograph was located
at 8,000 feet (2,440 m) in a deep canyon.

A comparison of the minimum temperatures for the stations at Tumamoc Hill showed the foot-of-the-hill temperatures to be consistently lower than at the Laboratory, halfway
up the hill. During the driest and most cloudless months of the
year, the difference was greatest, and only half as much during
the rainy months of midsummer. Shreve thought the reason
for the smaller difference in summer was due to the wetness
of the soil, which affects specific heat, and frequent cloudiness
at night, which affects the rate of radiation. The greatest difference between any of the daily minimum readings for the
Tumamoc Hill Laboratory site and the garden was 24°F
(13.3°C), and the least was 0.5°F (0.27°C) during the night after
a fall of 0.98 inch (2.45 cm) of rain.

During that summer several readings of the minimum for
the same night were made in mountains, or ridges and valleys
1,000 feet (300 m) below. These were taken in the humid
months and the readings differed from 4.5° to 8°F (2.5° to
4.5°C), less than the difference between the two laboratory
sites. In late September a comparative set of temperature readings was taken on the floor of the canyon at 5,000-feet (1,500-
m) elevation and on the side of the canyon 100 feet (30 m)
above the floor. These readings were alternately taken beginning at five minutes before sunset in the canyon. There was an
abrupt change of temperature at a particular level, the level
becoming higher as the evening advanced. On the floor of the
canyon was a stream of cooled air with a definite surface. The
greatest temperature difference due to cold air drainage was
nearly the same as the average shown by the two Desert Laboratory stations during September.

On the basis of his studies, Shreve concluded that the
influence of the cold drainage might affect both the upper limit
of lowland species and the downward occurrence of mountain

species. The downward limitation of forest and chaparral vegetation of the desert mountain range, however, is due to the
operation of the factors of soil and atmospheric aridity and not
to the cold air streams. On the other hand, the limitation of the
upper distribution of desert species appears to be a result of
cold air movements. He saw that a number of the most conspicuous desert species ranged from much higher altitudes on
ridges and higher slopes of canyons than they did in the bottoms and lower slopes of canyons. Soil samples indicated that
there was no essential difference between soil moisture on the
ridges and bottoms of canyons during the driest portion of the
year. The absence of the desert species from canyon bottoms
and their occurrence in higher elevations and ridges could be
attributed to the operation of cold air movements rather than
to the factors of soil and atmospheric moisture. Shreve concluded that an analysis of the operation of cold air drainage
plays an important role in determining not only the minimum
temperature but, more importantly, the duration of low temperature conditions.

Inversions of temperature are a common phenomenon
throughout the Sonoran Desert. In 1938 Turnage and Hinckley
used the term "ground inversion" to denote an atmospheric
condition in which the temperature increases with the elevation from the ground in basin localities to as much as 1,000 feet
(300 m) in the air. Above this zone a decrease of temperature
with elevation exists. The frequency of intensity and depth of
inversion varies with different meteorological and topographic
conditions. These inversions are essentially layers of cold air
that lie near the ground on calm, clear nights. They are highly
local in character, developed on nearly cloudless nights, require
only a short time to form, are quickly destroyed (as layers with
sharp vertical temperature gradients) by moderate wind movement and vary in numerous details under different conditions
of topography.

Turnage and Hinckley found that the hill station at the
Laboratory headquarters, 330 feet (100 m) above the plain was
near the top of the inversion layer. The garden station was on
the level plain about 100 yards (90 m) from the base of the hill.
These two stations are half a mile (0.8 km) apart and about 5
miles (8 km) from the Tucson Weather Station—at that time

located on the University of Arizona campus. The hill station was more representative of the general temperature conditions of the largest portion of the desert at elevations between 2,000 and 3,000 feet (600 and 900 m), while the garden station more closely approximated the conditions found in the basin areas, canyons, and valleys. They observed that freezing weather in the Sonoran Desert was the result of either the development of the ground inversion at night (which affects only certain areas) or the influx of cold northern air masses or both of these factors occurring simultaneously.

## Freezing and Plant Distribution

Low temperatures of winter had been recognized for a long time as an important factor in limiting the northward and vertical distribution of tropical and subtropical plants. According to Shreve (1911b), these warm-climate plants vary greatly among themselves in resistance to cold and different phases of winter cold that are fatal to them, and as a result there is a lack of coincidence in the northern limit of distribution among any considerable number of them. *The line that marks the extreme southern limit of frost is the most important climatic boundary in restricting the northward extension of perennial tropical species,* and it is the line at which the influence of winter cold is simplest in its operation. Subtropical species that extend northward from this line are subject to the limitation of a variety of aspects of winter cold. The principal ones are (1) the greatest number of consecutive hours in which the temperature falls below freezing, (2) the number of hours of frost in a single winter, and (3) the absolute minimum reached in the length of winter reckoned from the first frost of autumn to the last one in spring. For all subtropical plants the greatest number of consecutive hours of frost and the absolute minimum are the most important of these factors.

Tropical forms may be killed by continued exposure to temperatures just above freezing, and other forms may survive temperatures slightly below freezing but succumb under the formation of ice in their tissues. Certain plants will withstand the formation of ice in their intercellular spaces at a few degrees below freezing but will die at a sudden pronounced lowering of the temperature while they are in a frozen state. There are many arctic species, however, that will stand both freezing and very low temperatures. In many cases a sudden thawing proves fatal after a given set of cold conditions, although a gradual rise in the temperature enables plants to survive.

The winter of 1909–10 was the most severe winter in western North America for several decades, and Shreve chose it to study the relation of freezing to the distribution of saguaro and other plants of warmer regions. To obtain information on the greatest number of consecutive hours of freezing, Shreve used climatic records for four localities in Arizona in which thermograph records were readily available. These were Yuma, Phoenix, and Tucson within the range of saguaro distribution, and Flagstaff considerably above it. Passing from lower to higher latitudes or altitudes, the number of consecutive hours of freezing becomes gradually greater until the point is reached at which days without a midday thaw are first encountered. Then there is a sudden rise from about twenty-two hours to thirty-six to forty-two hours of frost. Shreve believed that the line in which this change takes place in the most severe winter is an important limit to plant distribution.

On the basis of figures available from the winter of 1909–10, it was found that the saguaro withstood nineteen hours of continuous freezing temperatures with a low minimum of 17°F (–8.3°C). Shreve concluded that the limiting factor in the distribution of saguaro is whether or not there is a warm period between two cold nights. He felt quite certain that there had in all probability been no days in Tucson in many hundreds of years in which the air temperature remained below freezing all day.

Shreve concluded that the occurrence of a single day without midday thawing, coupled with cloudiness that would prevent the internal temperature of the cactus from going above that of the air, would spell the destruction of saguaro.

To test this theory Shreve inserted thermometers into young saguaros subjected to various freezing conditions. Plants that were placed in low temperatures in excess of twenty hours in all cases became black and soft within six weeks and never recovered.

Shreve (1911b) tested resistance to freezing using

branches of staghorn cholla, a plant that extends in the desert mountain ranges to elevations to 5,800 feet (1,740 m) and the small form of hedgehog cactus which reaches 7,800 feet (2,340 m), the highest altitude attained by any cactus in southern Arizona. Both of these species were given various lengths of exposure up to sixty-six continuous hours—about the length that frost would supervene when two consecutive days were without thaw, conditions that probably exist in the mountains of Arizona at 7,800-foot (2,340-m) altitude in every severe winter. The plants experienced no injuries or fatalities.

Shreve (1914d) reviewed findings thus far on various phases of temperature factors limiting the distribution of plants and on the relative abundance of plants in different parts of their areas. There were two important considerations: first, those phases of temperature having to do with the length of the season in which growth and other activities were possible; and second, temperature conditions which would be a deterrent to the activities of the plant. The temperature phases of the growing season and of the frost season are not respectively reciprocal or complementary to each other. The end effects of temperature conditions of summer and those of winter are quite distinct, as they affect the limitation of the range of species. Quite commonly, it is the winter phase of temperature that limits the northern distribution and the summer phase that restricts them at their southern edge.

Turnage and Hinckley (1938) observed that two features of cold temperature stand out above others. First is the occurrence of frost; second is the duration of freezing temperatures throughout a night, the following day, and the following night, and perhaps longer. The southward limit of the first cold front coincides roughly with the northward extension of the tropical thorn forest (southern limit of the desert). The freezing duration coincides with the northern limit of the Sonoran Desert and with the vertical limit of desert vegetation on mountain slopes and tablelands. Although the most profound changes in vegetation occur at these boundaries, there are many species that have their distributional limits somewhere between them.

These observations corroborated Shreve's conclusions. In the Sonoran Desert there was no record of a single day (between two cold nights) when the air temperature failed to rise above 32°F (0°C). At Tucson there had been two occasions of nineteen consecutive freezing hours. The lowest maximum daytime air temperature at the Tumamoc Hill Station was 35°F (1.67°C) on January 7, 1913, and the second lowest of 36°F (2.2°C) occurred on January 26, 1937, when there were only two hours above freezing during the day preceded by nineteen hours and followed by fifteen hours below freezing.

Information concerning the various aspects of cold weather in the Sonoran Desert was imperfect, but it was believed that the thin layer of cold air near the ground, presumably at most topographic sites in inversion nights, was a potent factor in a seedling's struggle for establishment. Available temperature records seldom reveal information concerning duration of frost. In fact, very few records exist of soil temperatures and the duration and depth of soil freeze. Most observers have been content with a record of daily maximum and minimum air temperatures obtained at an arbitrary height and place and of relatively insignificant mean figures. In conclusion Turnage and Hinckley stated:

*Distribution of species under present climatic conditions surely has not reached a static state. Perhaps the northern limit of many Sonoran Desert species might be determined by factors other than winter cold, for instance moisture relations. Accessory physical conditions might greatly modify the influence of cold weather, to such an extent that a single datum of freeze is a poor index of the limiting factors when applied to different localities and at different times for the same species.*

## Conditions in Sun and Shade

A great deal has been written concerning the climatic interrelationships between plants. One important aspect that has been considered is the relationship between the overstory or tree canopy and the herbaceous or low woody plants. In some cases an antagonistic relationship has been reported, and in others more favorable conditions seem to have been set up as the result of the shade from trees.

Shreve (1931d) noted that throughout the arid parts of the United States and Mexico the shade of the large bushes and desert trees supports a much larger number of herbaceous plants, root perennials, and seedlings than do the unshaded areas alongside them. In southern Arizona, not only are the herbaceous annuals more abundant under tree shade—especially in the summer rainy season—but they reach a greater size than elsewhere and continue their vegetative activity longer than they do in the open. The shade is no more favorable than the open for the germination of seeds but is far more favorable for these plants during the critical rainy periods of their early history. Shreve pointed out that the paloverde and saguaro frequently grow in close juxtaposition and that in all such cases the paloverde is older than the saguaro. He thought this relationship was due to better conditions in the shade of the paloverde as well as the protection from mechanical injury that the tree afforded.

In order to determine the different effects of sun and shade, a study was set up with two stations, one in the open and the other in the shade of a foothill paloverde. The stations were located 23 feet (7 m) apart, on a nearly level outwash slope on the grounds of the Desert Laboratory. The tree had a height of 14 feet (4.3 m), a spread of 22 by 19 feet (6.7 by 5.8 m) with branches that hung low within 39 to 55 inches (1 to 1.4 m) of the ground. The shade was light and continuous except for a few patches around the periphery in the early morning and late afternoon. The leaves of the paloverde that appeared in the rainy season and persisted for two to six weeks were so small as to make little difference in the shade cast by the tree.

Weekly readings of absolute maximum and minimum temperatures were made, records of soil temperatures were obtained at depths of 3 inches (7.5 cm) and 12 inches (30 cm), and measurements of evaporation were made with Livingston's spherical atmometer in an open area and in shade. The precipitation data used were those obtained at the Desert Laboratory 4,300 feet (1,290 m) distant. The records were continued from January through October.

Under desert conditions the light shade of the tree supported a heavier stand of herbaceous annuals and seedling perennials than found in the open. The weekly totals of evaporation were very nearly the same in shade and open throughout the growing season. The moisture of the soil at 6 inches (15 cm) was usually greater in the shade, but the difference between sun and shade was only significant after very heavy summer rains or lighter rains in the cooler months. There were important differences in soil temperature both at 3 and 12 inches (7.5 and 30 cm).

Shreve concluded that the relatively close correspondence between evaporation and soil moisture in sun and shade indicated that the vegetational differences between these habitats were perhaps not greatly dependent upon differences in moisture conditions. The differences in solar radiation in the two habitats were not only responsible for the important differences in soil temperature conditions but also for differences in the water loss of plants, which were known to be inadequately indicated by atmometer reading of evaporation. (The intricate interrelationship of moisture and temperature conditions was also concerned in the fact that the deeper roots of plants were warmest at the same time of day that their branches and leaves were coolest, which must serve materially to aid the restoration of moisture content which the plant undergoes at night.)

## Evapotranspiration

During the period 1906–21, when Livingston was particularly interested in transpiration and evapotranspiration, there were very few data available on evaporation in the United States. Some work had been done—mostly by indirect determination based on humidity and temperature—but there were very few records from evaporation pans and some of these were of doubtful value.

Climatic factors that generally determine whether a given kind of plant may or may not live in a certain locality are divided into two groups, according to Livingston (1913b). The first group includes the factors that tend to increase or decrease the moisture content of the plant body. They may be termed the moisture conditions of the environment. The second group includes the climatic factors that tend to lower or raise the temperatures of the plant. If a plant form is observed as thriving year after year, generation after generation, in a certain locality, it is assumed that the plant is adapted to the environmental conditions. However, if the given form is not observed in this locality the only way to determine whether or not it can thrive there is to make experimental tests. The ability of a plant to thrive under arid conditions is often indicated by its observable physical structure. Its ability to withstand unfavorable temperature conditions, however—quite unlike its ability to withstand adverse conditions of moisture—is not as well indicated by structural characteristics.

Livingston pointed out that the moisture conditions of the environment affect the activity of plants by influencing either an increase or decrease of water content. Most plants derive water mainly from the soil and lose it mainly through the air; therefore, data on precipitation should supply a valuable criterion for comparing climatic areas. Livingston recognized the difficulties in the measurement of the evaporative power of the air and the many difficulties in the standardization of attempts to derive a formula by which evaporation might be computed from meteorological factors usually measured. These difficulties have since been overcome and the evaporating power of the air indirectly determined. Because there were no adequate methods of measuring the power of aerial surroundings to extract water from the plant at that time, he developed a porous porcelain atmometer to measure evaporation for local comparison with transpiration and for classifying geographical areas based on evaporation rates.*

This atmometer could be read daily, weekly, or monthly (in fact, at any convenient interval), the only condition being that the water supply in the reservoir must be adequate for the chosen period. Freezing weather and the effect of rain presented serious problems. The first was not subject to solution, but satisfactory correcting devices to prevent rain from entering the container were eventually developed by the users of atmometers.

*As the atmometer is not widely used at the present time and has been criticized by scientists interested in the measurement of evaporation, the discussion on the Livingston Atmometer has been somewhat curtailed; those interested in the development of the atmometer may find detailed information in the publications listed in the bibliography. R.F. Daubenmire (1974), however, considers the atmometer a useful tool in ecological investigation and mentions where atmometers may be obtained.

Livingston (1915a) took exception to the idea that free water somehow furnished the true standard evaporating power of the air. This proposition was probably related to a desire on the part of meteorologists and climatologists to compare rainfall with evaporation.

Rainfall is measured in depth units, and its measurement does not present serious difficulties; the size, shape, and material of the gauges exert negligible influence upon the readings, and the variation in exposure commonly allowed is not great enough to be important. Although it may be thought that evaporation—which is obviously a sort of negative rainfall—should be measured in the same kind of units, it has been found that many conditions influence the rate of loss from an evaporation measuring device and that the same conditions do not occur in each locality.

The secondary effect of evaporation in determining the moisture content of the soil was perhaps a more important factor in plant distribution than the primary effect of evaporation rate upon the plant itself. In any event, the optimum condition for growth was considered to be a balanced relation between the supply of water that the soil can furnish and the evaporating power there. Furthermore, the climatic evaporation is influenced by temperature, humidity, and wind velocity, all of which are very efficient in the control of plant activities. On this basis Livingston (1908c) was of the opinion that evaporation could be expected to play an important role in the determination of the great regions of plant distribution as well as in the distribution of plant communities within the same region.

In order to test this relationship, atmometers were provided to various observers scattered through the United States, and, of these, the data from sixteen stations were sufficient for the months June, July, August, and September to be used for comparative purposes.

The results obtained from the season's records were only partially satisfactory, but on the basis of the records that he did receive, Livingston gave a summary of the evaporation at various points in the United States. These ranged in order of decreasing evaporation and as follows: Mecca, California; Laramie, Wyoming; Tucson, Arizona; Salt Lake City, Utah;

Austin, Texas; Raleigh, North Carolina; Eugene, Oregon; Gainesville, Florida; St. Louis, Missouri; Burlington, Vermont; Lincoln, Nebraska; Chicago, Illinois; Orono, Maine; Newark, Delaware; Grand Rapids, Michigan; New York City, New York.

He correlated these records with the vegetation and noted the average for the northeast forest centers was the lowest of all with a figure of 93—the average weekly evaporation rate for the region for the northeastern conifers. The forest center showed evaporation of 107—the average evaporation for the deciduous forest center. The desert center of the Southwest had an average rate of 266.

This study was repeated the following year with a greater number of observers. Steps were taken to maintain a better standardization of atmometers, which were installed in pairs to make sure that the data obtained at these various stations would show reliable readings in comparison with one another. In all, thirty-eight cooperators were involved with the majority of stations in operation from May 25 to September 7, 1908. Based on the data obtained, Livingston prepared several charts and a map of the average weekly evaporation for the United States which together showed that evaporation rates were very highly correlated with the differences in vegetation. Again it was a case of the southwestern desert having the highest rate of evaporation; and the lowest rates of evaporation were in the Northeast and the Pacific Northwest. Livingston concluded:

*From the above general deductions it is apparent that the summer evaporation intensity alone furnishes a climatic criterion for studying the different vegetation centers with which we have to deal at least as promising as a criterion furnished by any other meteorological element. [Livingston, 1911d]*

Livingston believed that for the purpose of vegetational-climatic investigations, the controlling climatic factors are primarily effective only during the growing season and that the most promising approach was to adopt the length of the frostless season as the basis for comparison—that is, the days intervening between the average data of the last killing frost in spring and the first in autumn. These data are available from the United States Weather Bureau.

# Desert Soils and Physiographic Conditions

THE ORIGIN, MODE OF FORMATION, and development of desert soils have many features in common with all soils, but they are also subject to the distinctive actions related to deficiencies of rainfall or its intermittent and torrential character.

The basic importance of the underlying rock in determining the character of soils may be greater in the desert than elsewhere. The arid climate exerts itself in weathering of rock and the disintegration of rock fragments. In addition to the usual agents of disintegration, the effect of sand erosion is widely visible in desert areas. The results may be sand-carved rocks or sand dunes.

The removal of the finest soil material by wind and water erosion leaves rock fragments of various sizes, larger near mountains and smaller on the lower bajadas. On the more level area the results may be a cover of small pebbles, often referred to as desert pavement.

Desert soils are notably low in organic matter, a natural consequence of scant vegetation. They also tend to be highly mineralized, because the small amount of precipitation is not sufficient to leach the salts out of the soil. This may result in saline or alkaline soils. Under certain circumstances where lime is abundant it may result in the formation of rock-like lime hard-pan deposits locally called caliche.

## Sandy Areas

At the time of the establishment of the Desert Laboratory, the popular conception of desert among Americans was based on the traditional impression of the Sahara and Arabia, much more than on acquaintance with the arid regions of our own Southwest. It was widely believed that all deserts were sandy and were full of immense dunes kept in ceaseless motion by the wind. Although these characteristics may exist in the great desert areas in Africa and Asia, nearly all of the North American desert has a hard or stony surface, the only extensive stretches of sand being around the head of the Gulf of California.

Even in Egypt, MacDougal reported (1913b) that sand as a feature of desert landscapes is not so prominent as ordinarily supposed. One may travel for days without seeing either a mound or the unbroken stretch of sand that figures so largely in literature. In other places, a series of dunes a few miles in width may be found. These are known to extend in north and south directions for nearly 300 miles (480 km).

Although sand is not primarily a desert product, it is a component of the surface layers in most desert regions. Its presence may be due to one or more of several causes: for instance, denudation of a preexistent soil body and concentration of granular matter upon the denuded surface through removal of the finer components by wind action; existence of relict sand bodies of former geological conditions; direct deposition of volcanic sand in present or former geological times; or, to a lesser degree, the direct weathering and erosion of rock surfaces of suitable composition.

Generally, however, when great concentrated bodies of sand occur within arid areas, their presence denotes the operation of two contributing conditions in some adjacent or associated region: (1) conditions favorable for sand production upon a large scale and (2) wind circulation of sufficient intensity and fixity of direction to transport the material to areas where deposition and accumulation take place.

The two most potent agencies for production are combined wave and current action upon coastlines, and the erosive, transporting, and delivering processes of great river systems.

Sand, from either or both of such sources, may appear as coastal dunes or beach deposits that act as feeders for inland accumulation (Sykes 1939a).

One well-known sandy area lies in the Otero Basin in southern New Mexico within the White Sands National Monument. At the southern end the sand is a very clean white and is almost pure gypsum, supporting a sparse vegetation including several gypsophilous species. In northern Chihuahua, there is a large area known as the Samalayuca Sands in which there are some very large dunes. This area was visited by Coville and MacDougal and described in publications 6 and 99 of the Carnegie Institution of Washington (Coville and MacDougal 1903; MacDougal 1908b). The vegetation is similar to that of the White Sands.

Nearly one-seventh of the Lower Colorado Valley is occupied by sandy plains or dunes. The largest areas lie around the head of the Gulf of California and the Salton Basin; smaller areas lie west of the Mohawk Mountains and southeast of Parker, Arizona. The origin of the sand is not definitely known. It seems probable that it was largely derived from the broad beaches exposed by wide fluctuations of tide in the Gulf, and from the floor of the Salton Basin. Some of it is of local origin and has perhaps been derived from the granitic mountains lying within or bordering this subdivision.

Shreve provided a great amount of detailed information based on examination in 1936 of the sandy areas in Sonora (Shreve 1937c, 1938). The area covered lies in the northwestern corner of Mexico in the state of Sonora. At the time of Shreve's visit, this area of 6,670 square miles (17,342 square km) lying west of the 113th meridian had only two settlements with a total population of about forty. There was no agriculture, there were no cattle ranches, and on its longest road there was a stretch of 160 miles (256 km) without habitation or water. Since then, there has been much development and a paved highway borders the area on the north. This region is located in the driest portion of the North American continent with a precipitation of only 1 to 4 inches (2.5 to 10.0 cm) per year. Most of the rain is in the winter, but occasionally there are heavy summer cloudbursts. Little rain, rare frost, almost continuous sunshine, and three or four months in which the temperature goes above 100°F (38°C) are the earmarks of the truly desert climate that characterizes the entire sandy region near the head of the Gulf.

Shreve (1937c) described the Sonoran sands as a nearly level plain, but about one-sixth of the area was covered with active or stabilized dunes and about the same fraction with mountains. The surface of the plain was light gray in color, slightly crusted on the surface or covered with dark brown lichens, only rarely showing any of the angular bits of stone that commonly cover the surface of the desert.

In such an extreme kind of desert there is very little chance for plants to lead even a precarious existence. However, while the vegetation is indeed sparse, it is no more so than on the stony deserts of western Arizona and along the eastern edge of the Mojave Desert in California. Few species of woody plants, however, are able to live in these sands. Over the driest part of the region there were only three species of perennial plants, and there were extensive areas in which only a single perennial was found. Nowhere else in North America is the vegetation reduced to such simple terms.

The most striking feature of the sand was the contrast between the paucity of perennials and the great wealth of herbaceous plants that appeared right after a rain. These included a number of species that are rare outside the sands, as well as many that are found on other types of soil.

Paloverde, mesquite, creosote bush, and joint fir are the only long-lived woody plants that are common on the dunes. Where the dunes are very active, they support no vegetation at all, but wherever the movement of their surface has quieted down, they soon become colonized by big galleta grass and white bursage.

A great deal of the diversity in the large plants of the less arid desert is reflected in the small ones found on the sands. Usually each of the ephemerals stands apart and to itself. It is rare for the branches of plants to be interwoven, except in the wild buckwheats or the low sand verbenas. Where the plants grow together they are usually all of the same species and make a distinct patch in the varied pattern of the vegetation.

Some of the ephemerals are erect, as spanish needle and chaenactis; others form rosettes of leaves, as evening primrose;

low mosaics of foliage, as coldenia; or compact balls of leaves and flowers, as combseed. The size and form of the leaves vary greatly from species to species, and there is every shade of green from a deep rich tone to a light, almost gray color. In the Pinacate region and on the nearest mountains around the edge of the sand there are eight species of cacti that reach considerable abundance, but on the dunes and sandy plains there is scarcely one cactus to the square mile (2.6 square km). Examination of the cactus roots shows a long extension horizontally and very near to the surface, but sand is no place for a shallow rooted plant, for the surface dries out so quickly that there is only a brief period in which water could be absorbed. In heavier soils (i.e., mountains), however, they are thus able to absorb water from a moderate rain even if it does not penetrate far into the soil.

The greatest development of dunes has taken place west of Pinacate Peak, both along the Gulf coast on Adair Bay and in the interior. A belt of dunes, crossed by the highway between Yuma and El Centro, known as "the algodones" extends for 55 miles (88 km) along the eastern edge of the Salton Basin. There are innumerable small dunes along the west side of the Salton Basin and along the west coast of the Gulf in Baja California. They are often crescentic in shape and are gradually moving toward the northeast. Along the eastern edge of the sand in Sonora there are a number of lone dune areas projecting eastward over the gravelly plains, indicating the extent to which prevailing westerly winds have been responsible for spreading the sand.

About four-fifths of the sandy area has a stabilized surface and is relatively level, without local dunes. In these sandy plains there is little evidence of surface drainage and some evidence of gradual leveling from a less stable condition. Stabilization of the dunes is chiefly due to the coarse grass, big galleta, which is not usually abundant on the already stabilized sandy plains. Neither galleta nor shrubs have been as important in binding the level surfaces as have the blue-green algae and ground lichens that form a nearly continuous crust on the most stable plains.

Active or moderately active dunes are either devoid of plants or have a very irregular stand of perennials. When sand accumulates around the plant slowly enough for growth to keep pace with it, water supply is improved. When sand is removed, the plant frequently perishes or persists only because of a deep-seated root system. The only perennials commonly found on moderately active dunes are creosote bush, white bursage, Mexican tea, white dalea, saltbush, coldenia, desert buckwheat, and Thurber sandpaper plant.

Rare examples of other perennials of the surrounding region may be found on unstable sand, including the blue paloverde, burro brush, and desert willow. Characteristic of the dunes is the prevailing dryness of the surface in contrast with the relatively good moisture supply of lower levels. The absence of cacti from deep sand is significant in this connection. The occurrences of big galleta indicate that it only becomes established on sandy surfaces that have already passed the stage of most active movement. When this stage is reached, big galleta, white bursage, white dalea, and coldenia increase greatly in abundance, and the stabilization of the surface is rapidly accelerated.

## Desert Pavement

An interesting and easily observable surfaces oil condition is that in which the so-called "desert pavement" is formed. Although such pavements may vary somewhat in character and in some localities may be composed of different materials, they are structurally much alike. They consist essentially of expanses of otherwise bare ground covered with a mosaic of pebbles—small fragments of stone lying so closely together that it is difficult to detect the loamy soil in which they are embedded. These stones are nearly uniform in size and appearance, and their upper surfaces are frequently darkened and polished by wind-scouring and weathering processes that produce the finish known as "desert varnish."

The environmental conditions necessary for the development of pavements varying in size from a few square yards to thousands of acres are an available supply of pebbles in a level, or almost level, area of ground isolated from surface drainage channels, and an air circulation of sufficient intensity to remove the finer components of the top soil.

Sykes believed that the formation was largely due to the

action of wind erosion but that sheet flooding during a storm is the agency by which the conspicuous leveling is accomplished and may sometimes account for the development of the desert pavement.*

## Caliche

One of the common characteristics of arid soils is the development of a calcium carbonate layer known in western Northern America as "caliche." Caliche is a mixture of colloidal clay and carbonated lime. These two constituents are carried from the surface of the desert soils to the depth to which occasional rains penetrate at a depth of a few inches to 3 or 4 feet (1 or 1.2 m). At the level where the wetted soils dry out, a more or less compact caliche stratum is formed (Spalding, 1909c).

Shreve and Mallery (1933) made an extensive study of caliche in the Tucson area. They observed that caliche was abundant in the soils of outwash plains, or *bajadas*. The heaviest deposits were found in the lower parts that had been built up totally or largely by outwash from volcanic mountains. Caliche was also found in the soils of limestone and volcanic hills covering the submerged rock surfaces with filled-in crevices. Generally, it was absent from purely granitic soils, from sand, from heavy clay of alluvial floodplains, and from similar but somewhat coarser alluvial soil of poorly drained playas. The uppermost layers of caliche were observed in some localities very close to the surface and at depths as great as 6 to 10 feet (2 to 3 m) in other localities. They occurred near the surface only in places that had been subject to recent erosion.

The surface of caliche is relatively smooth and undulating. The thickness of the layers varies from a fraction of an

---

*The sources from which the pebbles have been originally derived and the processes by which they become nearly uniform in size and shape are in most cases rather obscure. Many observers believe that the development of paved surfaces has taken place through the interoperation of wind and occasional rain-flooding in eroding and removing earlier superimposed strata in which the stones have been embedded, but recent studies indicate that the pavement may result from migration upward of small stones in the surface soil.

inch to 16 inches (a few millimeters to 40 cm) but is most commonly between one and two inches (2.5 and 5.0 cm). The lateral extent of the caliche layers is extremely variable, and the continuity of the surface is frequently broken. The upper surface layer is free of stones and soil particles. The lower portion is somewhat softer without visible structure and filled with a greater or lesser number of stones of various sizes. The soil beneath caliche appears to be normal but often is highly impregnated with salts giving a white appearance. It usually contains pebbles or nodules of material closely similar to layers of caliche. At increasing depths, successive layers of caliche are sometimes similar to the surface layer; in other places, somewhat softer and less sharply defined. Recurring layers have been found in numerous well-digging operations in the large valleys of southern Arizona that have been filled with detrital material at great depths. Under the eastern part of the Tucson area, they have been found in close succession to a depth of 100 feet (30 m).

Penetration or rise of water in loam soils is very materially retarded by layers of caliche from ½ to 1 inch (1.25 to 2.50 cm) thick. It is only by reason of discontinuous extension of the caliche layers that water is able to penetrate the soil. On the other hand, moisture present in bodies of soils overlaid by one or more layers of caliche is retained much more effectively than in soil at the same depth and texture that was without caliche. The roots of shrubs and trees are able to penetrate the caliche through the cracks or around the edges of the layer only by chance. The character of the root system of a particular species determines its success in reaching the lower layers of conserved water. The superficial root systems of cacti rarely penetrate the uppermost layers of caliche. Deeply rooted trees, such as mesquite, are infrequent and small in soils with several layers of caliche. On the other hand, creosote bush with its roots richly distributed in every direction is the most successful plant in penetrating soils filled with caliche and is the commonest plant on such soils in the Tucson region. Areas without caliche in the Tucson region support the greatest number of trees and shrubs which reach their greatest abundance in the upper *bajadas* of the Santa Catalina, Tortolita, Santa Rita, and Rincon Mountains—where the mineralogic character of the surrounding mountains are such that there has not been a heavy accumulation of carbonates in the soil and, consequently, there is a restricted formation of caliche.

## Soil Studies

Detailed soil investigations at the Desert Laboratory began early and were continued throughout the next thirty-eight years. They included a variety of investigations: soil moisture, soil temperature, the physical and chemical characteristics of soils, and the interrelation between climate and soils and vegetation and soils. These soil studies are of great general interest since they help us to understand the environment in which desert plants have survived, and even flourished.

Plants depend upon favorable moisture and temperature conditions in the soil, and often the ability of animals to survive in a hot arid environment is also dependent upon belowground conditions. Air temperatures may be relatively similar over wide areas, but soil temperatures vary widely because of differences in exposure to the sun's rays. These factors affect soil moisture, resulting in marked local microclimate differences.

### Moisture

Livingston (1906c) was the first of the Desert Laboratory scientists to become seriously interested in the soil moisture conditions in the Desert Laboratory area. He was surprised to find what he considered to be a great amount of moisture, sometimes amounting to 10–20 percent by volume between rock fragments at depths of 16 inches (40 cm) after periods of four to five weeks without rain.

In 1905, which happened to be an unusually wet year, he collected soil samples to determine the amount of moisture in the soil prior to the summer rainy season. As of July 1, the desert conditions on the hill were nearing their maximum severity for the year. Surface soil around the Laboratory building was air-dry and seemed thoroughly baked. Day and night temperatures varied from 80° to 105°F (26° to 41°C) or above, and the daytime relative humidity varied between 8 and 15 percent. No rain had fallen since May 12, at which time 0.77 inch (1.97 cm) was measured. The only plants in good condition were those particularly adapted in some manner to the low moisture content of the soil. In rocky soils, the moisture content of fine material between rocks could have been below the wilting percentage, but there may have been available moisture in pockets and on the surface of rocks and caliche. The amount of moisture in the Tumamoc Hill soils at the end of the early summer dry season was more than Livingston had expected (1906c). His explanation was that the surface layers of the soil are dry during most the year. After a shower they dry out rapidly and, in so doing, shrink in such a way as to be somewhat loosely porous to a depth of several inches. The deep cracks characteristically produced in many heavy soils upon drying from a puddled condition are not found. Although cracks often form, they are small and close together and seldom penetrate more than a few inches below the surface.

The high evaporation demand of the desert air removes water from these surface layers much more rapidly than it can be replaced from below, and this soon results in an air-dry condition near the surface. The evaporating surface retreats farther and farther into the soil, and evaporation is hindered more and more by the thickness of the nearly air-dry layer through which the water vapor must diffuse upward. Eventually, an equilibrium is reached in the rock-bound pockets of Tumamoc Hill at a depth of less than 3 feet (1 m), as is shown by the actual amounts of water noted in the dry season, and possibly also by the position of the caliche layer that may roughly mark the position of the average evaporating surface throughout many centuries. Thus, as he reasoned, the surprisingly large amounts of water found near the soil surface even at the end of the dry season are due to the presence of a thick layer of air-dry soil.

Livingston thought that the lack of an adequate water supply may be effective in two ways. First, the actual moisture content of the soil may be too low; and, second, there may be sufficient moisture in the soil to supply the plants for many days, yet the plants may suffer because the rate of movement of this water may be inadequate to supply the soil layers immediately surrounding the roots as fast as these layers are exhausted by absorption. These two conditions are closely related and difficult to separate. Also, as the water is removed from the soil, the concentration of the soil solution may increase so that, as the critical point is approached, it becomes somewhat difficult to distinguish between actual paucity of water and the effects of high osmotic pressure.

He explained that the ability of plants to survive under the moisture conditions previously discussed was in part due to the characteristics of the plants themselves. For example, seedlings of ocotillo roots often penetrated to a depth of 4 inches (10 cm) or more within forty-eight hours after the first appearance of the "seed leaves" and therefore would be able to make use of a very brief moisture situation in the surface soils. The creosote bush has much the same habits in germination. Because the deeper layers of soil dry out very slowly after rain, it might be seen how such seedlings, germinating in the rainy season, might reach to a depth where they would have a permanent and adequate water supply before the upper layers of the soil had dried out sufficiently to result in death.

Livingston conducted studies to determine the minimum water supply for desert plant survival. Because it was difficult, if not impossible, to make accurate measurements of transpiration and water supply of plants growing in the ground, small plants were grown in tin cylinders, perforated at the bottom to facilitate drainage. Condensed-cream cans, holding from 10 to 12 ounces (296 to 355 cc), were found to serve admirably for this purpose. Because of the voracity of the desert animals—insects, birds, and small mammals—it was necessary to protect the cultures by wire netting, but this cover did not materially affect the water relations of the protected plants.

The plants chosen for this experimentation included some extreme xerophytes and others more mesophytic in nature. One problem was digging up the mature specimens of desert plants that live through the dry seasons. Their roots penetrate far into the soil through openings between the rock fragments, and they cannot be removed without injury. Even with smaller forms such as spurge, the leaves and stems of which extend upward for several inches from the summit of a long, woody primary root, it was seldom possible to excavate deeply enough to discover all the lateral roots.

Some of the plants used were native ephemerals that spring up everywhere after the first rains. These plants are not especially xeromorphic; they appear to be very much like the smaller annuals of more humid regions. Although they reach a height of only a few inches above ground, they develop a root 8 inches (20 cm) long following the rains and soon penetrate to such a great depth that they are not injured by the rapid and almost complete drying of the surface soil. Several cultivated plants of humid regions were grown from seed and used as a comparison.

Livingston found very little difference between germination percentages of plants from humid regions and those from the desert. The advantage that the desert forms have over those of humid regions is the rapid growth of the tap root, which reaches the deeper soil layers and allows the desert plants continued growth after the surface soil has dried out.

Livingston said that without doubt the open formation of desert vegetation makes it possible for the plants to draw upon a very large volume of soil for their water supply. He thought that the scarcity of seedlings and young plants, even in the rainy season, could be explained in part by the fact that conditions other than those of available moisture are effective in reducing the number of these individuals even though in most years, the surface layers of the soil do not remain moist long enough after each shower to allow the seedlings to obtain a foothold. He believed it to be more probable that the animal life depredations are the most important factor in preventing the growth of seedlings. The young plants, exposed in great numbers early in the growing season, provide a possible source of water for animals and may be consumed in large numbers. This observation was shared by others, notably MacDougal and Shreve.

Up to 1912, Livingston, and in fact most soil scientists, had given little attention to the variation in amounts of available moisture as related to soil texture. It was also assumed that, because xerophytes were more able to resist wilting than crop plants and most mesophytes, they could survive under lower soil moisture contents. There was much surprise and disbelief following the 1912 publications of Briggs and Shantz* who, in a series of papers dealing with the problem as one of soil physics rather than of plant physiology, announced conclusions in sharp contrast to those of earlier workers.

Working with about twenty soils that differed widely in water-holding capacity and other physical characteristics, and employing more than one hundred species and varieties of plants ranging from typical hydrophytes to such extreme xerophytes as the cacti, these investigators found that the moisture content of any particular soil was practically a constant at the time the rooted plants had become strongly wilted. The authors concluded that this "wilting coefficient" for a given soil was a fixed quantity for all species grown in it, at all stages of their development, and that differences in environmental conditions during development and wilting were without effect upon its magnitude. The amount of moisture in the soil at which wilting took place was dependent upon the texture of the soil itself. For example, clay had a relatively high percentage of moisture remaining in the soil at the time of permanent wilting, whereas sandy soils had a very low percentage of moisture in the soil. The amount of moisture in the soil was also found to be related to the total capacity of the soil to hold moisture. That is, the clay soils had the highest retentive capacity and, at the same time, the highest wilting coefficient, whereas the sandy soils had the lowest water-holding capacity and, at the same time, the lowest wilting coefficient. Briggs and Shantz brought out another relationship at this time that changed the thinking of soil scientists. It had been the custom to saturate a portion of soil, allow it to drain, and then weigh the soil, dry it, and weigh it again to determine the amount of moisture held by the soil. This was considered to be the water-holding capacity. Briggs and Shantz, however, found that in the field this was not a true situation. They reported that a soil becoming saturated immediately following rain or irrigation

---

*Homer L. Shantz, a distinguished botanist and plant geographer, was president of the University of Arizona from 1928 to 1936.

would lose water by percolation until an equilibrium was reached. This equilibrium might be much below the so-called water-holding capacity. Briggs and Shantz termed this particular moisture relation the "field capacity."

The botanists at the Desert Laboratory believed that xerophytes had a particular ability to withstand low soil moisture levels without permanent wilting. They pointed out that it is obvious that a cabbage plant or bean plant will not grow side by side with the xerophytes under arid conditions, but that the crop plants will wilt and die much sooner than the desert plants. While making his Death Valley studies, Coville (1893) had stated that "If a plant of ordinary structure such, for example, as red clover were exposed to the climate conditions of the desert; it would wilt, dry up and die."

Livingston had previously found, by growing plants in tin cans, that xerophytes would maintain life long after crop plants had wilted under the same conditions of soil moisture (Livingston, 1906c). Partly because of these observations and partly to determine for themselves the actual conditions under which plants wilted and died in a desert environment, the scientists at the Desert Laboratory undertook a series of experiments more or less duplicating the work of Briggs and Shantz. The main difference between the two studies was that Briggs and Shantz did their work in a greenhouse, whereas the Desert Laboratory scientists conducted most of their work under field conditions. The results of their studies were reported by Brown (1912), Caldwell (1913), and Shive and Livingston (1914).

The experiments included evaporative conditions varying from shade to direct sunlight. Plants tested included devil's claw, cutleaf groundcherry, nasturtium, cocklebur, corn, and kidney beans. Their conclusions were that while the general soil moisture relationships reported by Briggs and Shantz held true under ordinary evaporation conditions, there was a difference in favor of the desert plants in the ability to survive under severe desert conditions of high evaporation rates.*

---

*Part of the difficulty is that no method has been developed to measure the moisture available to plants under field conditions, where moisture may remain in cracks in rocks for a long time.

## Temperature

Investigations were made at the Desert Laboratory to determine surface soil temperatures and temperatures at various depths for all seasons of the year. In addition, Laboratory scientists made investigations to determine the effect of soil color on soil temperature, the effect of soil moisture such as irrigation on soil temperatures, and the effect of aspect and exposure, not only in the lower areas of the desert, but extending up the Santa Catalina Mountains to elevations above 8,000 feet (2,400 m).

Temperatures of the soil were important to both plant and animal life, as they relate to the environment in which roots are growing in various seasons of the year and to living conditions in rodent burrows. Small mammals and some other animals are able to endure the high temperatures of the desert because they are largely nocturnal and spend the warmer parts of the day in burrows of a lower temperature than the air above the surface.

It was known that in a humid climate in western Europe, the soils at 6 feet (2 m) have an almost constant temperature throughout the year. In contrast to this, the records for the desert conditions indicated an annual fluctuation of 19°F (10.5°C).

The daily June surface temperature in the Sonoran Desert reached a maximum of 165°F (73.9°C). At 3 feet (1 m) there was no daily range but a slight weekly one. At 6 feet (2 m) there was a very slight weekly variation and at 12 feet (4 m) there was none. Annual fluctuations at 6 feet (2 m) were 19°F (10.5°C). Annual extreme temperatures at 12 feet (4 m) lagged from three to five months behind the extremes for the air, with an annual fluctuation of 7.5°F (4.2°C).

The difference between minimum soil temperatures and minimum air temperatures was greatest during the coldest periods of winter. On cloudy nights the two were approximately equal. During the winter, the maximum soil temperature was seldom more than 10°F (5.5°C) above the air maximum, but in March the difference began to increase rapidly. There was little seasonal change between the relations to the minima; the maximum surface soil temperature rose during May and June to levels as high as 165°F (74°C), about 50°F (27.5°C) above the maximum air temperature. The surface soil warmed rapidly after sunrise, increasing about 90°F (50°C) to the maximum about 2:00 P.M. By 3:00 P.M., cooling set in strongly and by 8:00 P.M. the temperature had fallen 50° to 60°F (27.5° to 33 °C) below the 3:00 P.M. reading. The rapid changes are related to the dry atmosphere and the lack of heavy vegetation cover in the desert regions.

One of the early studies at the Laboratory was an investigation of the effect of color on the temperature of soil surface. In this study three soil thermographs were located 4 feet (1.2 m) from each other at a depth of 3 inches (7.5 cm) in a brown clay of basaltic origin. After it had been determined that the temperatures of each of these three areas was approximately the same, the area approximately 4 feet square (1.5 m²) above one of the thermograph bulbs was then blackened by application of drought black dissolved in turpentine, giving a uniform black matte to the soil and small rock particles. Over the second bulb the soil was covered with a very thin layer of crushed mortar so as to provide a uniform white without any minute reflective surfaces. Over the third bulb, which served as a control, the soil remained in the natural light brown color. During the succeeding four weeks of the experiment, the days were mostly clear with two partly cloudy days. During these four weeks the average maximum of the black soil was 2.8°F (1.5°C) above the control and the average of the white soil was 7.7°F (4.3°C) below that of the control, making a difference of 10.5°F (5.8°C) between the black and white soils. The average minima in both cases were nearly identical. These studies brought out the fact that natural desert soils are only about 3°F (1.5°C) cooler than a blackened surface soil that absorbs the maximum of heat rays. Also, the difference between the superficial temperatures of natural soils that are nearly white and those that are dark would never be greater under these desert conditions than 10.5°F (5.8°C).

In order to find out what happens when the soils become moist in the summer rains, a study was undertaken to determine the effect of artificially applied moisture. A spot was selected with fine, adobe soil on the level floodplain. The moisture of the soil was raised by sprinkling the surface with a watering pot, using measured quantities of water (simulated rainfall), and applying it uniformly over the areas during a period of twenty to fifty minutes. The amount applied was equivalent to either a quarter (0.63 cm) or a half-inch (1.25 cm) of rainfall.

The complete deformation of the normal daily temperature curve took place in all cases only on the day that the water was applied. On the following day the curve was normal in form but reached its daily maximum thirty to sixty minutes later than the control. There was a marked progressive increase in the daily maximum temperature for five to six days. On the following week there was still an effect on the soil temperature, although the moisture of the soil had returned so nearly to the condition of the control that the customary sampling method determined no appreciable difference. The most prolonged effect followed a half-inch (1.25 cm) wetting with a lowering of 11°F (6°C) below the control twelve days afterward.

In the situation studied, the lowering of soil temperature by irrigation is in great measure due to the cooling effect of the evaporation of moisture from the surface of the soil. Under climatic conditions in which evaporation is not so active as in the desert, the cooling effect of wetting is not so marked, and in northern latitudes it may happen that early spring rains will cause an increase in the soil temperature, as reported in the office of Dry Land Agriculture of the United States Department of Agriculture.

Shreve set up an interesting study designed to determine the effects of exposure on temperature. He had a small hill of fine alluvial clay constructed to a height of 10 feet (3 m) with a ridge running east and west, sloping at an inclination of 30°. After allowing a year for the hill to become settled, three soil thermographs were installed at a depth of 3 inches (7.5 cm), one on the north slope, one on the south slope, and one on the level ground. At the end of the arid foresummer, when the moisture content of all three areas had been reduced to approximately the same level, it was found that the south slope was from 9° to 13°F (5° to 7.2°C) hotter than the north slope. The average weekly maxima for the south slope and the level ground were nearly the same during April, but beginning with the second week of May, the maxima for the level ground began to become greater than for the south slope. Shreve's explanation for this phenomenon was that, with the approach of the summer solstice, the sun's rays fell more directly on level ground than on a slope of 30° southern inclination. Based on rather extensive studies of the relation of slope exposure to soil temperature, Shreve (1924c) pointed out that the influence of slope exposure in modifying the character of vegetation outside tropical latitudes indicates that the angle of incidence of the sun's rays is the fundamental determinant of the environmental differences between opposed slopes that face north and south. He believed that the temperature of the soil is the immediate factor by which differences of insolation affect the other aspects of the environment, such as the rate of water loss, the warming of the lowest layers of the atmosphere, and the lengthening or shortening of the growing season.

Shreve (1924c) made studies during a two-year period on soil temperatures as influenced by altitude and slope. During 1921, over a period of five weeks from early April to the middle of May, he obtained readings by soil thermographs at depths of 3 inches (7.5 cm) and at altitudes of 3,000 feet (900 m), 4,000 feet (1,200 m), and 5,000 (1,500 m) on north and south slopes in the foothills of the Santa Catalina Mountains, near Tucson. The soil in each case was closely similar granitic loam and the slopes were approximately 15° from the horizontal. He was surprised to find that maximum soil temperature increased with altitude. Also, throughout the period, the maximum readings for south-facing slopes were slightly lower than those of the north-facing slopes. The only possible explanation for this appears to be that the late afternoon sun falls on the north-facing slope more directly than on the south-facing slope and, in the absence of plant covering, causes the already warm soil to attain a slightly higher temperature in the afternoon.

# Desert Plants
# and Their Relationship
# to the Environment

## General Environmental Relationships

*The behavior of any plant is controlled by surrounding conditions—varia-
tions in these being the stimuli, or causes, which produce in the internal,
physiological complex of the organism various responses or effects. Such
responses are, however, as dependent upon the nature of the responding
organism as upon the nature of the stimuli. Within the same set of environmental conditions, different plants behave differently—merely because their
internal conditions differ—and in unlike environmental complexes, plants of
the same form exhibit different behaviors. The behavior of plants thus depends
upon two interacting sets of conditions, one being external and the other
internal. The latter set makes up the nature of the plant and serves to define
it physiologically. These internal factors determine the ability of the plant
to respond to any given constant external condition, or to any given change
in any condition, and also determine the extent of such responses. [Livingston
and Shreve, 1921]*

The scientists at the Desert Laboratory defined vegetation
as the total plant population of an area viewed from the
anatomical and physiological rather than the taxonomic and
floristic standpoint. Livingston and Shreve approached questions on plant distribution with the idea that they are merely

146

problems in physiology. All the environmental conditions are fluctuating and uncontrolled, but nevertheless measurable, and all the activities of the plant are in normal performance and measurable by such features as distributional extent, habitat occurrence, communal behavior, relative abundance, size, and seasonal behavior.

The physical conditions of the desert resemble a sieve or screen, with meshes of a certain magnitude, through which will pass only those successful plant forms that withstand the most adverse conditions of the environment. Furthermore, the size of the meshes in the screen is continually changing throughout the year, while the size of the imaginary particles that are to be screened is also undergoing continuous change with the advance of the organism from phase to phase of its development.

## Plant Response to Available Moisture

Water relations hold center stage in plant physiology, and it might truly be said that the importance of this subject was the basis for the establishment of the Desert Laboratory.

In various experiments, a number of saguaros, barrel cacti, and various opuntias were removed from the desert soil, then placed on suitable supports in the accustomed upright position—thus, lengthening the dry seasons to which they had been subject in nature. Some of the test plants were placed in the open air, some in the more equitable conditions of a well-lighted laboratory room, a few were kept for periods of a few months in a constant temperature darkroom, and others were exposed to the full blaze of the Arizona sunlight, standing on a base of black volcanic rock—thereby avoiding none of the desiccating effects of the climate.

The water balance of a plant and the length of life which an individual might survive at the expense of its great accumulated supply, and what activity it may carry on while cut off from the customary supply of soil moisture, were the objects of several studies that showed wide variation among the species considered (MacDougal, 1915a). The saguaro may live a year or even two upon a balance in the Arizona open. Growth and reproduction are in the main inhibited by any notable depletion; however, the death of the main trunk of a plant may leave a living branch held high in the air, and this may produce flowers.

MacDougal, Long, and Brown (1915) reported on the results of desiccation and starvation of succulent plants. For example, a barrel cactus may lose as much as 1/2,000 part of its weight in one day, immediately following its removal. The same plant six years later, under equivalent conditions—except that its weight had been reduced nearly a third—lost no more than 1/17,000 part of its weight in one day. In the sixth year, the loss was 1/20 of the water supply at the beginning of that year.

The authors concluded from various studies that green plants, when subjected to relatively great diurnal evaporation intensity, frequently exhibit a marked fall of leaf moisture content by day and a corresponding rise by night.

The greater part of water absorbed by growing plants merely makes up for the loss occasioned by transpiration. The daily water loss from any individual plant at any stage of its growth is largely determined by the evaporating power of the air. During the development of the plant and under the corresponding conditions of soil moisture, there is a maximum rate of water supply to the transpiring tissues; when the evaporating power of the air is so great as to cause the rate of loss to surpass that of supply, the plant wilts and is injured or killed, unless it is provided with water storage organs that may temporarily free it from dependence upon external supply. In most plants, the quiescent existence can be maintained when the rate of transpiration approaches or equals that of supply, but under these conditions very little or no growth can take place.

Livingston (1913d) made a rather detailed study of the resistance of leaves to transpirational water loss. He noted that the rate of water loss from plant surfaces exposed to the air was always lower than the rate of evaporation from an equal area of water surface in the same surroundings. In other words, evaporation rate is usually much greater than transpiration rate. It is thus possible to regard plant tissues, especially in many desert plants, as exerting a retarding influence upon evaporation and outward diffusion of their contained water.

This slowing effect on transpirational water loss was of different magnitude for different plant forms, for the same form grown under different conditions, and for different portions of the transpiring surface of the same individual plant. Furthermore, for the same plant, and for the same portion of its surface, the resistance to transpiration often varied with the age of the organism and with the diurnal march of various internal conditions. The factors that condition the retarding influence are all to be characterized as internal, that is, they are all operative within the plant body, though they are ultimately controlled by external conditions. These factors are anatomical in that they depend on the nature, amount, and arrangement within the plant body of its various component substances. For example, the resistance to water loss by transpiration is much greater in leaves having a well-developed cuticle than in others in which the cuticle layer is thin or absent. Other internal conditions being equal, a leaf with closed stomatal pores must exhibit a greater retarding influence upon water loss than one with open pores.

Leafy plants uniformly showed a high transpiration rate during the day and a low during the night; however, the cacti exhibit the opposite, having a high period in the night and a low one in the day. Livingston concluded that there exists in cacti a mechanism for governing the rate of water loss that is entirely different in its response to external conditions or in the daily periodicity from the corresponding mechanism of leafy plants. This was later confirmed by Edith Shreve (1915).*

Spalding believed that the amount of transpiration in some xerophytes was in direct relation to the amount of water available in the soil in which the plant was growing. He grew seedlings of creosote bush, blue paloverde, and Mexican paloverde (horsebean) in cans supplied with water at stated intervals. He found, for example, that the rate of transpiration of blue paloverde and Mexican paloverde decreased to 52.6 and

---

*More recently it has been found that many succulent plants are able to store carbon dioxide resulting from respiration at night and use this as a source of supply for photosynthesis during daylight hours. They do not need to have open stomata to obtain carbon dioxide from the air, which is the case in most green plants.

38.5 percent, as high as it was on the previous day, as a result
of diminished water supply. In an experiment with creosote
bush, the results were more striking. Two plants, grown in
small containers, were weighed in the forenoon; one was given
three ounces of water at noon, and the other received no water.
In the afternoon, it was found that the plant receiving water
was transpiring three times as rapidly as before, and the other
was transpiring only one-fifth as rapidly. While these studies
were far from exact, he concluded that the rate of transpiration
is affected by the availability of water in the soil (Spalding,
1905b).

## Cell Sap Concentration (Osmotic Values)

Heinrich Walter, a scientist from Europe, visited the
Desert Laboratory and became interested in the forces by
which the plant could obtain moisture from soil.

He had discovered earlier that the osmotic strength of cell
sap serves as an indicator of hydration of the protoplasm on
which he believed depended the course of all physiological
functions of the plant—such as growth, photosynthesis, and
respiration. Therefore, the determination of osmotic values in
the sap of a plant serves as a basis for the understanding of its
water relations (Carnegie Yearbook 29).

During six months' residence at the Desert Laboratory,
Walter determined the osmotic values for a number of the
characteristic plants of the desert vegetation, as well as for
plants in the evergreen oak region and the mountain forests of
the vicinity. From October 1929 to April 1930, a total of more
than 1,000 determinations of osmotic value were made.

For each species of plant, there are definite and characteristic minimum, optimum, and maximum osmotic values. The
osmotic value is lowest during the intensive period of growth;
then, in the fully developed leaves it reaches a value higher
than may be regarded as the optimum one. Under unfavorable
conditions of water supply, the concentration of the cell sap
rises still higher. When a definite maximum value is exceeded,
the leaves die. Plants of the same ecological type are characterized by similar osmotic values and similar relations between
the critical points.

Walter stated that the lower the osmotic value the more intensive is the growth. It follows, conversely, that under favorable conditions for growth, the osmotic value is relatively low. Accordingly, clear relationships are found between the distribution of a plant species and the osmotic value of its sap. In the center of the distributional area of a species, the osmotic value is always relatively low. On proceeding from the center toward the distributional limits, it gradually increases and finally reaches its maximum value. This holds true for distributional limits controlled by water relations as well as for those determined by temperature conditons.

It was believed by Walter and other scientists that the osmotic values indicated the ability of the plant to extract moisture from the soil and that the concentration of cell sap was just as important in ecological and plant geography investigations as the peculiarities of the external and internal plant structures.

## Root Systems

It is the common habit, when referring to relation of "plant" to its environment, to mean the aerial portion only, leaving quite out of consideration the subterranean part. The obvious reason for this seems to be that roots do not greatly excite our admiration or curiosity and thus have received little attention in the field. Further, relatively little experimental work has been done on the roots of plants—other than on seedlings and the growing of plants in solution. Soils and the soil conditions are more difficult to express in a manner capable of ready application than is the aerial environment of the plant (Cannon, 1916).

There were many obstacles to desert plant root investigation—notably the rocky substrate in which most are found. W. A. Cannon, however, overcame these difficulties and provided invaluable information on the underground parts of desert plants that have been so often overlooked in the more superficial studies.

He once remarked that if desert plants could be up-ended with their root system above ground, the desert cover would be so complete as to resemble that of a tropical rainforest—an entirely different picture than that afforded by the widely spaced plants we see on the desert surface.

Cannon observed that desert plant roots are remarkable for their individuality. The roots of each genus, often perhaps of each species, possess peculiarities of form, branching habits, color, texture, position in the ground, and physiological reactions. He rejected the popular idea that roots of desert plants are very long—that is, that they penetrate the ground to great depths in search of moisture and that the length of root systems is a measure of the aridity of a locality. Moisture at great depth is only found in the bottom lands where water tables are developed from percolation of water from adjacent or distant slopes. In uplands, moisture never penetrates more than a few feet (few decimeters), and roots are largely restricted to the moistened soil layers.

## Types of Roots

Cannon grouped root systems into three general types: one characterized by lateral growth; one by growth in development of the tap root; and one by a generalized root system, that has both the tap root and laterals well developed. This last type is typical of desert plants such as creosote bush, triangle leaf bursage, and mesquite. Most of the annual plants have this form. The species with a tap root as a chief feature include joint fir and spiny allthorn. The lateral growth type is typified by the cacti which are almost the sole representatives (Cannon, 1911).

Specialized root systems are changed little with environment, but the generalized roots are often extremely variable, ranging from a pronounced tap root to a marked development of the laterals, dependent on soil characters and water relations. Each of the species has its own characteristic of branching habit, although this is a matter of degree rather than anything else. For example, the roots of creosote bush branch repeatedly wherever the plant is growing, but those of ocotillo are little branched. The most richly branched root system observed was that of pencil cholla, which covered the ground so completely that it would have been impossible for any other plant to gain a foothold without encountering the roots of this species.

Perennials with the generalized type of root system have the widest local distribution, and those with a pronounced development of the tap root have the most limited distribution. Plants with well-developed laterals, the cacti especially, are most abundant on the bajada and on Tumamoc Hill (where the soil is shallow) and seldom occur on the floodplain.

Neighboring plants exhibit a certain degree of competition. The roots of annuals intermingle and often occupy the same horizon. Where perennials of a single species occur together on the *bajada*, the roots of one plant may reach to and intrude upon the root area of its neighbor. In studying creosote bush on the *bajada*, sixty roots of neighboring creosote bushes were encountered that either were in contact with the roots of the plant studied or were in the same horizon. On the floodplain, competition among the roots was often not as great as on the *bajada*. The roots of annuals growing on the *bajada* reach as deep as most perennials in the same habitat, and, since they occur in large numbers, competition with them must be an important causal factor contributing to the sparseness of the perennial vegetation of the *bajada*. The annuals also come into competition with the shallow rooted perennials through the laterals that are developed on the tap root of the annuals— 1½ to 2 inches (4 to 5 cm) beneath the surface of the ground.

Cannon pointed out that the root systems of winter annuals are usually distinguishable from those of the summer annuals. The most striking characteristics of the roots of the former are the prominently developed tap root and the meager development of the laterals that are generally filamentous or at least thin. Summer annuals, on the other hand, have root systems frequently of more generalized form; that is, the laterals are well developed and are frequently rather coarse, and the main root is often forked. The absorbing surface of the summer annuals appears to be greater than that of the winter forms, although the depth of penetration of the roots of the two classes are apparently about the same.

The causes for differentiation in the root systems of the winter and summer annuals are not totally known, but they may lie in the nature of the species as well as in the different environments of the two classes of plants. The rains of summer, in addition to thoroughly wetting the soil, serve to cool it as soon as they come. The air temperature immediately falls, and the relative humidity at once becomes high. In brief, the conditions for a tropical luxuriance of growth are at hand, and the shoots of the plants appearing at this season bear a great number of large leaves; the transpiration surface is relatively great.

In the winter season, however, the general character of the annuals and the character of their environment are strikingly different from the conditions just described. The temperature of the soil begins to fall with the coming of the summer rains and continues to decline until March-April. Therefore, the rains of the winter season do not materially change the course of the curve as do summer rains. Air temperature during the winter rains is relatively high, but much lower than summer. There are also occasional periods of really high temperatures and drying winds. The winter plants usually do not grow as quickly and do not have as large a leaf surface as the summer annuals.

Given a sufficient amount of water both in winter and summer, it is possible that the difference in the relative temperature relations of soil and air for the two seasons is the main operating cause in bringing about the difference in the root systems. In summer, the temperatures both of soil and of air permit very rapid growth and the best conditions for water absorption—with the result that the root system of a plant is well developed. In winter, on the other hand, the soil at the 8- to 12-inch (20- to 30-cm) depth is in daytime colder than the air. This operates directly to depress the rate of water absorption, to limit the development of the roots, and thus to make the conditions unfavorable for the fullest growth of the shoot, with the result that the demands on the root system are relatively low. Were the winter soil warmer than the air, the growth of the shoot would probably be much more vigorous and the root systems of the plant much more extensive and of a different character than they actually are.

Cannon (1912b) followed the seasonal activity of root development throughout the year to learn about relationships of roots to moisture and the soil-air environment. Examination of desert plant root systems during the season of drought failed to show portions with vegetative activity, although it was possible at the same time to demonstrate a certain, even if low, rate of transpiration. In winter many of the plants native to the Southwest do not form new roots, in spite of the fact that some plants, particularly the chollas and prickly pears, absorb water very promptly after rains. Cannon concluded that temperature and moisture and in many cases the aeration of the soil were requisites for root development.

## Root Systems in Vegetation Distribution

The most arid portions of the deserts of the southwestern United States are on the higher lands and the less arid portions on the lower lands—the floodplains or the washes. It is only in the less arid areas that plants with pronounced tap roots occur. In the more arid regions fleshy plants are almost entirely absent and root systems characteristic of such plants are consequently not found.

It was Cannon's belief, after studies in Algeria as well as in southwestern United States, that, if any type of a root is entitled to be called xerophytic, it is the generalized form and not the deeply penetrating tap root. He made a distributional comparison of the three types of root systems in the Desert Laboratory area: (1) those having a generalized root system with both laterals and tap root well developed; (2) those with the tap root as the most prominent feature; and (3) those with the laterals of special prominence (Cannon, 1911).

Among plants with generalized root systems, creosote bush, catclaw, hackberry, brittlebush, paloverde, and mesquite may be found to some extent in each habitat, both where the soil is deep and where it is shallow. With these forms should also be included the annuals except those with bulbous roots. Plants with this root type are either evergreen or deciduous, but all have fairly large transpiring surfaces, and in some of them, especially brittlebush, the leaves are large.

Plants having a prominent tap root include condalia, night-blooming cereus, joint fir, spiny allthorn, and graythorn. These forms are restricted to localities—preferably the floodplain—where the soil has sufficient depth for the development of the main root.

The cacti are nearly all provided with the third form of

roots, but to them should be added sangre de drago and krameria—the latter a parasite. The cacti occur both on the *bajada*
and on Tumamoc Hill, but not on the floodplain of the Santa
Cruz or near-by West Wash to any extent. They are the plants
par excellence of the *bajada*.

## Root Responses to Soil Temperature

Cannon pointed out, however, and it is generally recognized, that soil acts as a reservoir for heat and the daily course
of soil temperature is unlike that of the air immediately above
it. Thus, roots are subject to temperature conditions that are
quite different from those affecting the shoot of the same organism. The shoot is warmer by day and colder by night than
the root. The roots of most cacti of the Tucson region and
possibly elsewhere lie near the surface of the ground. For the
most part they are less than 12 inches (30 cm) deep. As the rate
of root growth of cacti is relatively slow at temperatures much
under the "optimum," the importance of the shallow positions
of roots is apparent—it is only in the upper horizon that such
favorable temperatures are to be found. On the other hand, the
deeply placed root system such as mesquite may have a relatively rapid rate of growth at relatively low temperatures and
the root response to temperature is much less important than
that played by the roots of cacti.

Cannon attributed the presence of cacti in the Tucson area
in large part to the occurrence of rains during the season when
the soil is also warm. This observation makes it possible to
suggest that a similar root-soil relation may be found among
the cacti of other regions, explaining on the one hand their
presence in such regions and suggesting on the other the causes
for their sparseness or absence in yet another.

Cannon (1917) reported on the study of the root growth
of seedlings of mesquite. These studies were made in glass
tubes filled with sand, using a nutrient solution or tap water,
mostly the latter. The tubes were kept under controlled temperatures to maintain soil within 0.9°F (0.5°C) during the
course of any experiment. Minimum root growth took place
about 54°F (12°C) with 108°F (42°C) as the maximum.

The effect of warm temperature on root growth is shown
in part by Coville (1893) who called attention to the enormous

length of mesquite roots in the dry bed of the Armagosa River between Salt Wells and Saratoga Springs in Death Valley. There he observed a 52-foot (15.8-m) root of a mesquite tree that had been washed out by a torrent. The diameter of the root at the large end was 1.1 inches (2.75 cm) and at the smallest end about 0.59 inch (1.5 cm), and there were no lateral branches.

Cannon gave particular attention to the root growth of creosote bush related to temperatures at a depth of 12 inches (30 cm) at the Coastal Laboratory in California and the Desert Laboratory in Tucson for the month of August 1916. At all temperatures, growth of the creosote bush at the Coastal Laboratory failed. He found that the optimum temperatures ranged from 59° to 90°F (15° to 32°C) and concluded that 90°F (32°C) is about the optimum temperature for the growth of roots of creosote bush (Cannon 1918a).

An important aspect of the soil temperatures is the great variation at depths of 6 to 12 inches (15 to 30 cm) in the amount of heat between winter and summer. At no time of the year is the temperature of the soil at a depth attained by the roots of mesquite so low as to inhibit root growth. In ocotillo and cholla, low temperature is possibly the most important factor by which it is limited.

Cannon related soil temperatures to a study of the needs of all aerobic plants for oxygen. The role of oxygen in the environment is in part influenced by the medium in which the plant lives as well as the morphological nature of the plant itself, and oxygen relations are fundamentally quite different between roots and shoots.

Roots in the same system that extend widely or penetrate deeply have unlike relations to oxygen, owing mainly to differences in the oxygen of different strata of the soil and the amount of oxygen in any stratum that may vary from time to time from different causes, both climatic and biologic.

The interrelation between temperature and oxygen content was found to be very important. Where the oxygen supply falls below the norm, the oxygen supply and not the temperature determines the growth rate of roots (Cannon, 1925b).

In all species, there was a concentration of oxygen in which growth would not continue whatever the temperature might be. He noted that creosote bush, ocotillo, and cholla, which are intolerant of poor conditions in the soil aeration, grow on the *bajada* slopes near Tucson. While the soil of the *bajada* is coarse and probably well aerated most of the year, during the rainy season it is frequently puddled. Even so, there may be translocation of oxygen in solution and in sufficient amount to promote root growth and water absorption by the plant. These species are largely absent on the floodplain where the soil is adobe and saturated during the period of high temperature and where lateral water movement is not active. Mesquite, a species common to the floodplain, is relatively insensitive to oxgen deprivation.

Cannon and Free (1917) found that increased air supply, if not excessive, favors root branching and probably accelerates the rate of root growth for cholla and ocotillo. They observed that the composition of the soil atmosphere was neither the same nor as constant as the composition of the atmosphere. The presence of living matter in the soil, including bacteria, fungi, and protozoa, as well as the roots of higher plants, tended to decrease the oxygen of the soil atmosphere and to increase its content of carbon dioxide. These, together with chemical reactions associated with the decay of dead organic material, tended to impoverish the oxygen supply of the soil and enrich the carbon dioxide supply—which must then be counteracted by the diffusion of the soil atmosphere and the general atmosphere, possibly assisted by changes in barometric pressure and in temperature of soil and air.

The importance of root habits in ecology has long been recognized on the basis of characteristic and specific reactions to features of the soil and environment included under temperature and water relations. It has been shown, for example, that the general distribution of the cacti is closely related to the response of the roots to the temperature of the soil. It seems probable that soil aeration must be added as a factor of no less importance than temperature and water.

## Desert Seedlings: The Establishment of Vegetation

The obvious differences of soil texture and topographic site determine the flora of a given area, but the relative abundance of the different types of plant species is not so much a matter of physical conditions as they affect the adult plants, but rather an outcome of the vicissitudes of germination and establishment for the species of that area.

Shreve believed that the makeup of vegetation and even, to a large extent, the density of the individual plant stands are products of specific conditions that control germination and activities of the seedling during its first twelve months (Shreve, 1911a).

His observations on the germination of desert perennials indicated that the hard-leaf species germinate at lower percentages of soil moisture than do the succulents. The development of the root system of the seedling and the changes in minimum water requirement with advancing age show that the atmospheric factors are quite as important as those of the soil in determining the behavior of seedlings. Striking differences were found in the manner of establishment of different classes of perennials, and survival was found to be due not so much to chance variation in the seedling as to small differences in the immediate environment of different individuals *(Carnegie Yearbook 9)*.

The characteristic openness of desert vegetation (in contrast with forest and grassland) affords ample room for the establishment of new individuals of the perennial species. The number of germinations that take place during any rainy season in the desert is very great, however, as compared with the establishments that result.

The variability of rainy seasons in the Southwest has much to do with the persistence or death of seedlings. Low water content of the soil often results in the death of the seedlings after the close of the rainy season. This may occur, even after several years of successful seedling growth, because the root systems have reached soil that is already being drained by long-established perennials.

The circumstance of many germinations but a subsequent high death rate for hard-leaf shrubs is comparable to the circumstance of succulent perennials that have few germinations but a higher percentage of survival, not only in the first, but in succeeding years. Deep-rooted plants that survive the first dry period are sometimes killed in the dry seasons of subsequent years, whereas the relatively shallow rooted seedlings of cacti are much more liable to die from the impact of heavy rains or the accompanying burrowing and scratching of rodents, or being eaten by animals in search of moist food.

On a 6,000-square-feet (550-square) plot located on the north slope of Tumamoc Hill, Shreve observed the reproduction for various species. In May 1910, the vegetation of this area included 24 creosote bushes, 22 foothill paloverdes, 10 whitethorns, 9 staghorn chollas, 6 ocotillos, 3 Toumey prickly pears, 3 lyciums, 2 saguaros, and 2 Kramerias.

The germination of all the perennials took place during the week following the first heavy summer rain. It commenced promptly, and growth proceeded with great rapidity. Shreve observed on the second morning that there was already evidence of germinating activity, and on the third morning it became possible to identify the young seedlings. In the study area there were two saguaros, and these produced many thousands of seeds every spring so that there should have been a sufficient supply to provide for reproduction. At the start of his observation, Shreve found a single seedling of saguaro; in 1913 this seedling was 0.6 inch (1.5 cm) high and 0.82 inch (2.1 cm) in diameter. In 1917 the seedling was still alive and had a height of only 1.7 inches (4.2 cm) and a diameter of 1.5 inches (3.8 cm). The slowness of this growth is characteristic of the saguaro.

During seven years of observation on this plot, not another seedling of saguaro was found; in all field work during these same years, Shreve found only four seedlings less than 2 inches (5 cm) high.

According to MacDougal (1908g) the seeds of the saguaro, which are produced in enormous quantities, are devoured in large part by birds even before being freed from the fruit. Of the great number that reach the ground and germinate, not more than one in a million survives to make the curious globular plantlet a few inches (5 to 10 cm) in height destined to become a great cactus.

Observations on the barrel cactus produced similar germinating data. In 1910, a single seedling was found; it reached a height of 1½ inches (3.8 cm) in 1913 and a diameter of 2.4 inches (6 cm). In 1917 this plant was 4.3 inches (10.7 cm) high and 4.7 inches (11.7 cm) in diameter. No other germinations of barrel cactus were located. As indicated above, the rate of growth for the barrel cactus is more rapid than that of the saguaro; however, the populations of the two have about equal density throughout the desert in general—although they are not always equally abundant in any one place (Shreve, 1917b).

The foothill paloverde showed an abundance of early germinations but the resulting plant growths were not many. Total number of germinations in nine years was 1,226; number of survivals was only 19. Heaviest mortality occurred in the two or three dry months immediately following the germinating season.

Shreve found that the germination rate of foothill paloverde was extremely variable, ranging from 542 in 1910 to none in 1915. This variability is due to the fact that seed crops are borne only in very favorable years and the waxy coating of the fresh seed delays germination from one to several years. The mortality rate in seedlings of the paloverde is determined by soil moisture conditions and not by competition with each other or with other vegetation.

The ocotillo is another plant that produces a copious crop of seed almost every year, and the seeds are so light as to be thoroughly distributed. The number of seedlings that appeared during the first days of germination was so great "as to make it utterly impracticable to keep track of them." Several hundreds might occur on almost any square meter of ground.

Mortality of the ocotillo seedlings immediately after rain was extremely high—not more than one in every 10,000 to 100,000 seedlings survived to the next summer. In 1910, there were 5 young plants of ocotillo on the area. In later years, 21 others were recorded; the largest number found in any one year was 5. In July 1917, only 7 plants were surviving. This represented a relatively high percentage of those that reached their second summer but an extremely small percentage of the total germinations.

The seedlings of all the cacti form a favorite food for a large number of small animals, being juicy reservoirs of water and containing enough other material to lead to their destruction before sufficient armament has been formed for their protection.

## Flowering and Fruiting

The flowering period of cacti in the vicinity of the Desert Laboratory commences early in April, at the close of the winter rainy season, and ends around the middle of December when the last flowers of the jumping cholla are seen.

Because flowering may occur during dry or rainy seasons, it appears that the production of flowers is not in any way dependent upon the abundance of moisture in the soil. Although there are some exceptions—the fishook mammillaria appears to wait until the summer rains come—many plants, particularly the barrel cactus, may flower abundantly for at least six months (Lloyd, 1907a).

Lloyd observed that the flowers of the organ-pipe cactus were open only at night, while those of a close relative, the saguaro, were open early in the morning and remained so throughout the daylight hours. Anatomically, the flowers of the two species are similar. It is interesting that the flowers of the desert night-blooming cereus appear quite tender, whereas those of the organ-pipe cactus do not—yet the sun never shines upon them.

Flowers of the saguaro are usually more numerous on the east side of the stem or more advanced in development. D. S. Johnson believed that the probable reason is a higher average temperature on the east of the thick stem, which is the side first warmed by the morning sun and which retains its temperature late into the afternoon *(Carnegie Yearbooks 11, 12, 13 and 14)*.

Plants that flower early in the morning (a little after sunrise) include several of the prickly pears and chollas—notably the cane cholla and staghorn cholla, the mammillarias, the saguaro, and hedgehog cactus. The large flowers of the Arizona rainbow cactus open about 7:00 A.M. and are fully expanded in half or three-quarters of an hour.

The flowering habits of the jumping cholla fascinated Lloyd. Someone could set his watch by observing this cactus flower and be scarcely fifteen minutes away from the true time. When he took out his watch at 2:30 P.M., not one flower was out; yet fifteen minutes later there were many flowers showing. He cited other observers who indicated that a flower on this plant never has been seen before 3:00 P.M. (by sun time). Later he modified this statement, saying that the jumping cholla may open at 2:30 P.M., a half-hour earlier, on cloudy days. The flowers remained open until 9:00 P.M. when the first signs of closure were observed. The pencil cholla has much the same habit.

Johnson also observed that in the cane cholla colors range from deep red to bright yellow within the species, although the variability on any one plant is usually narrow. Also, the flower of any individual cane cholla is constant from one season to another. The difference seems to rest upon the degree to which the deep red, always manifest in the apical portion of the petal, overspreads the remainder of the flower. These color relations in the flowers can be correlated with differences in structure of the flowers, fruits, and stems.

Reproduction comes from the fruits that serve as propagules for the development of new plants. If seeds happen to be present in the fruits, they are in no way affected by the sprouting of the fruit. An interesting development of fruits is that the axillary buds give rise to flowers as long as the fruits are attached to the mature plant, but when the buds fall, the fruits give rise to shoots and roots only (Johnson, 1918).

The jumping cholla matures a great number of greenish, pear-shaped fruits with apparently perfect seeds. Flowers open in May to July and mature their fruits [.78 x 1.18 inches (19.5 x 29.5 mm)] in late summer. An interesting characteristic of this fruit is its persistence on the parent plant for ten to twelve or perhaps even for fifteen to twenty years. During this time, the fruits continue to grow until they may become 1.5 inches (37.5 mm) in diameter and 2.75 or 3 inches (68.7 or 75 mm) in length. During every summer, four or possibly more generations of fruit may be added to the jumping cholla chains in a single growing season. Hence, the longest chains observed, made up of twelve to fourteen fruits, may have been formed in three or four years, or as many as ten. A few, or as many as fifteen to twenty of the dormant buds of this persistent fruit produce new flowers. The fruits of these behave in like manner, so that in ten to twelve years a parent fruit may have even hundreds of daughter fruits depending upon it.

## Acclimatization

*Every species inhabits the areas in which it has been able to reach and occupy from the starting place of its origin. Neither its birthplace or any of the places within its range may offer the most suitable conditions for the best growth and highest development. Beyond seas, over mountain ranges, across the equator or past other equally effective barriers may lie plains, valleys, plateaus, or even continents where if once introduced it might overbear all complication from the plants already there. Extending its distribution a thousand fold and the number of its individuals a million fold. Let the barriers be but once passed and it enters into a new kingdom as the various invasions of weeds testify. [MacDougal, 1907d]*

In 1906 plots were established at several locations in the vicinity of the Desert Laboratory, in the Catalina Mountains at elevations ranging from 6,000 to 8,000 feet (1,800 to 2,400 m) and, a few years later, at the Carmel Laboratory to test the ability of plants to survive in new environments.

The Carmel Laboratory near the shore of the Pacific Ocean has a cool, equable, foggy coastal climate. Temperatures from 1909 to 1920 ranged from 34° to 104°F (1° to 40°C), with the lowest temperatures in January and February and the maximum usually in September and October, although it was in May in some years. Rainfall ranging from 9.63 to 26.21 inches (241 to 655 mm) was restricted to winter months, mostly in December, January, and February. The soil was sandy with a low moisture-holding capability.

Plantings at the Desert Laboratory were made in two places; one near the main building at an elevation of 2,800 feet (840 m) and the other at the base of the hill at an elevation of 2,300 feet (690 m) in a place subjected to cold air drainage. The maximum temperature recorded was 113°F (45°C); minima were 10°F (–12°C) on the hill and 5°F (–15°C) in the lower area. The growing season varied from 275 to 290 days. Summer rainfall varied from 3 to 10 inches (7.5 to 25 cm), and the annual average was 12+ inches (30+ cm).

The Xero-Montane Plantation was located in the Santa Catalina Mountains, at an elevation of 5,600 feet (1,680 m) in the encinal or oak woodland. Temperatures ranged from 18.5° to 100°F (–7.5° to 38°C), and the growing season was 204 to

244 days. Summer rainfall recorded during the years 1908 to 1914 ranged from 6.05 to 12.21 inches (151 to 305 mm) with an average of 8.88 inches (222 mm).

The Montane Plantation at an elevation of 8,000 ft (2400 meters) furnished the equivalent of the forest climate in Michigan and New York, with a temperature range from 16° to 86°F (-9° to 30°C). The growing season endured from 138 to 153 days, and summer rainfall recorded in 1908-14 ranged from 11.40 to 27.82 inches (285 to 695 mm) with an average of 18.34 inches (458 mm).

Seeds and living plants of 139 species were introduced into these four areas; the total number of introductions was 192, and with repetitions, the operations totaled 250. Most transfers were from one area to another, but some plants came from eastern United States and one from Mexico. The behavior of these plants was observed for fourteen years, from 1906 to 1920.

The Montane Plantation showed the establishment and survival of 7 species out of 28; the Xero-Montane 7 out of 33, the Desert Laboratory 6 out of 21; and the Coastal Laboratory 41 out of 100.

The greatest percentage of survival—38 percent—was shown by plants transferred as resting aerial shoots; of the transfers by seed 30 percent survived; rhizomes and thickened roots gave 26 percent; and corms and bulbs showed only 10 percent.

Transfers from the Montane Zone to the coast had the largest proportion of survival, 15 out of 18 species. Of 56 transfers from the arid stations to the coast, only 14, 9 of which were cacti, survived. No species from the Atlantic seaboard or Pacific coastal region survived at the Desert Laboratory. No plant from the coastal region, from the Desert Laboratory or from the Xero-Montane station survived at the Montane plantation. Nearly half of the 28 species from the Atlantic seaboard introduced at the Coastal station survived.

Perennial forms native to the lowlands of eastern America, with a growing season of 168-70 days in the winter and extended exposure below the freezing point, endured the climate of the Montane Plantation in Arizona at 8,000 feet (2,400 m)—where the growing season is scarcely over 100 days—in which alterations of aridity and humidity are rapid and frequent and the winter conditions severe.

Species indigenous to the region of the Montane Plantation for the most part survived when taken to the foot of the mountain. Some species from the arid region survived in the equable maritime climate, but the fleshy succulents from the latter region quickly perished when taken to the desert, demonstrating the wide difference between the succulents of the desert (such as cacti) and the fleshy plants from the beaches. Spineless or nearly spineless cacti from Arizona developed and retained a decided armature when grown at the Maritime Plantation.

MacDougal observed that there were some unusual factors affecting the advance of prickly pears up dry mountain slopes. Chief among these is the part played by animals in modifying dissemination. An unarmed species of prickly pear spread through the oaks and up into the region of the pine. At higher levels it is found chiefly in the crevices of huge boulders and masses of rock little visited by rodents or other animals looking for food. However, the fruit of the prickly pear is attractive to birds. The entire distribution of the plant might be accounted for as carried from point to point by birds, and, as seeds matured even at the higher levels, it seemed destined to remain throughout its present range.

MacDougal reported that a half a dozen rooted species of cactus were placed in three spots in successive years near the vicinity of Barstow, California. Observations were to be made on the behavior of plants native to a desert—with summer as well as winter rainfall—when transferred to a region in which the major precipitation occurs in the cooler season. The plants were all quickly devoured by native rodents, and it was concluded that animals would constitute an effectual barrier against the movement of some Arizona plants into the Mojave Desert.

As another example, smooth prickly pear from the foothills below the Xero-Montane plantation, was taken to the Xero-Montane plantation in 1915; a second lot in 1916, and a third lot in 1917. All were eaten before the close of each season. A later introduction with the use of wire-screened cages was successful in this location; the lots outside the screen were eaten. On the basis of these observations it was concluded that the small animals native to the Xero-Montane region effectually limit the distribution of smooth prickly pear.

It is rarely possible to ascribe the stoppage or restriction of a species to the direct and simple action of a single physical agency. The actual effect of defective humidity, low or high temperatures, soil moisture, and other environic components are each modified in a manner by the intensity of degree of the others, together with an allowance for rapidity of variation in these factors as well as time or duration of exposure (MacDougal, 1907d).

## Longevity of Cells

MacDougal and Long (1927) made a detailed study of what takes place in cells of barrel cactus and the saguaro that have attained a very great age. Some of the cells of the barrel cactus seemed to persist in a functioning condition for periods of a century or more, after the manner of the saguaro. Although changes in the plant take place during these many years, the cells remain alive.

MacDougal discovered that the foothill paloverde had living cells after 250 years. Studies show the existence of ray cells of the thin wall type in the foothill paloverde, as well as typical tracheids with walls in parts of trunks formed over two and a half centuries ago. In earlier work it was shown that thin-walled cells of saguaro that have lived for several centuries may continue to enlarge for over half this period while similar elements in the barrel cactus showed capacity for growth during nineteen years of the century that they endure. The long-lived cells in other plants examined were all of the large thin-walled type. That wood cells or tracheids should remain alive and retain plump and normally appearing nuclei after so long was considered remarkable and without parallel in any previous observations.

# Plant Life Forms

THE MAJOR LIFE FORMS OF PLANTS found in the Sonoran Desert are shown in Table 7.1.

The vegetation of the Sonoran Desert is distinguished from that of adjacent regions by the characteristics of its dominant plants. They are different in height, bulk, manner and amount of branching, character of stem, size, duration, color, seasonal behavior of leaves, degree of succulence, and time of flowering. The appearance of the landscape is a result of the mixture of these life forms which may be simple—where rainfall is sparse—to very complex, where soil conditions are favorable and where rainfall reaches a maximum for the desert.

## Ephemerals

The desert environment offers particular advantages to the short-lived plants. The restriction of vegetative activity to brief moist periods almost completely relieves the ephemeral plant from the difficulties of water supply that beset the perennial. By speed of germination, rapidity of growth and early flowering, and maturity during favorable periods, the ephemeral escapes the most potent conditions controlling the life and ecological behavior of desert plants (Shreve, 1951; Shreve and Wiggins, 1964).

Ephemerals appear following periods of moderate or heavy rain, whether this rainfall occurs in the cool or hot months or at rare and sporadic intervals. There is a minimum of rainfall below which ephemerals do not appear, and a maximum that is a deterrent to them. They are scarcely affected by the annual or seasonal totals of rain, by the length and severity

## Table 7.1. Life Forms of the Sonoran Desert

| Life Form Types | Examples |
|---|---|
| Ephemerals | |
| Strictly seasonal | |
| Winter ephemerals | 1. Wild carrots, Indian wheat, fetid gold poppy |
| Summer ephemerals | 2. Pectis, tidestromia |
| Facultative perennials | 3. Desert marigold |
| Perennials | |
| Underground parts perennial | |
| Perennial roots | 4. Penstemon, anemone |
| Perennial bulbs | 5. Onion, ajo lily |
| Shoot base and root crown perennial | 6. Tobosa grass, three awn grass |
| Shoot perennial | |
| Shoot reduced (a caudex) | |
| Caudex short, all or mainly leafy | 7. Agave |
| Leaves succulent | |
| Leaves nonsucculent | 8. Beargrass, sotol |
| Caudex long, leafy at top | |
| Leaves simple, semisucculent | 9. Yucca |
| Leaves branched, nonsucculent | 10. Fan palm |
| Shoot elongated | |
| Plant succulent (soft) | |
| Leafless, stem succulent | 11. Barrel cactus |
| Shoot unbranched | |
| Shoot branched | |
| Shoot poorly branched | |
| Plant erect and tall | 12. Cardón, saguaro |
| Plant erect and low or semiprocumbent and low | 13. Mammillaria |
| Shoot richly branched | |
| Stem segments cylindrical | 14. Cholla |
| Stem segments flattened | 15. prickly pear |
| Leafy, stem not succulent | 16. Stonecrop |
| Plant nonsucculent (woody) | |
| Shoots without leaves, stems green | 17. Cruxifixion thorn |
| Shoots with leaves | |
| Low bushes, wood soft | 18. Brittlebrush, bursage |
| Shrubs and trees, wood hard | |
| Leaves perennial | 19. Creosote bush, jojoba |
| Leaves deciduous | |
| Leaves drought deciduous | |
| Stems specialized | |
| Stems indurated on surface | 20. Ocotillo |
| Stems enlarged at base | 21. Torote, boojum tree |
| Stems normal | |
| Stems not green | 22. Sangre de drago |
| Stems green | 23. Paloverde |
| Leaves winter deciduous | |
| Leaves large | 24. Cottonwood |
| Leaves small or compound | 25. Ironwood, mesquite |

of drought periods, by the high maximum temperatures of summer, or by the freezing temperatures of the winter.

The most favorable local habitats for ephemerals are merely those affording an abundant and briefly sustained moisture supply. Character of the soil is of little importance aside from its penetrability and retentiveness. Soil depth, structure, and mineralogical origin, all so important to perennials, mean little to the ephemeral. Sand, however, is particularly favorable for ephemerals of the cool season, because of the deep infiltration of moisture and the rapid surface warming early in the season and early in the day.

It would be difficult to make a local differentiation of habitats solely on the basis of short-lived herbaceous plants. Their distribution in the Sonoran Desert is determined by the moisture of the soil, the character of the soil surface, and the available shade of perennials. On sand, wind-blown ephemeral seeds are covered, held, and left at a favorable depth for germination. In general, the plant crop is more evenly distributed over a sandy soil than over any other.

On hard, smooth soil surfaces with little vegetation, there is rarely a thick crop of ephemerals, as their seeds are easily washed away or blown away. Any slight depression in such a surface is usually thickly colonized. On undisturbed desert pavement the chances for seed lodgment are poor, and few germinations take place.

Where small stones project from the soil, there is a light or moderately heavy crop of ephemerals. Localized but abundant stands appear on slopes covered with large stones. In the shade of perennial plants, ephemerals are larger and more numerous than in the open; under the low, spreading shrubs, the crop is larger than under trees. Shrubs and the accompanying dead undergrowth arrest wind-blown seeds and then serve to conserve moisture.

In slight depressions or on benches where conditions have been favorable for the accumulation of seeds at the end of the active season, the number of seedling ephemerals of one species may be very great. On a slightly shaded level area of deep, moist clay near Tucson, 1,200 seedlings of Indian wheat were counted on 10 square feet (1 square meter) (Shreve, 1951). Such colonies are probably due to the depositing of seeds by a rain runoff that occurred just when the seed crop of a predominant species was ripe for distribution.

Dense stands of ephemerals are common on floodplains and in other areas with level alluvial soil, a fact that points to floodwater as the agent in their distribution. Such stands are particularly characteristic of fetid marigold, Indian wheat, and careless weed. On gentle slopes and other surfaces covered with coarse gravel, the ephemerals are more widely spaced and numerous species more evenly intermingled.

In the Sonoran Desert there are two distinct groups of ephemeral species, one confined to the warm rainy period, the other to the cool.

The temperature for optimum seed germination in winter ephemerals is between 60° and 65°F (16° and 18°C), and for summer ephemerals between 80° and 90°F (27° and 32°C). The narrow margin between the highest germination temperature for winter ephemerals and the lowest for summer ephemerals is sufficient for sharp seasonal separation of the two groups. In winter the seeds of summer ephemerals remain dormant under favorable conditions of soil moisture; in summer the seeds of winter ephemerals also lie in a moist soil without germinating.

Little is known about the period for which viability is retained by desert ephemeral seeds. It has been observed that an exceptionally favorable season will bring forth large crops of species that have been uncommon for ten to fifteen years or more. Also, the relative abundance from year to year of the regularly recurring species does not bear a constant relation to the extent of the population and seed crop of the immediately preceding two or three years.

In southern California native herbaceous plants are perforce limited to the winter and early spring, since the summer is rainless. There are almost one thousand species of annuals, and about half of these are native to the Mojave Desert and adjacent areas. If the ephemerals of the rich flora of the Mojave Desert are followed eastward, their numbers are found to diminish rapidly through the Colorado Valley and the Arizona Upland. If followed south along the Gulf Coast, they are found again to diminish on both sides of the Gulf. Both to the east and to the south a few species not found in the Mojave Desert are added to the group.

## Winter Ephemerals

Winter ephemerals are abundant throughout the Sonoran Desert, but the number of species decreases toward the northern and northeastern limits. They are most conspicuous on the sandy plains of the Lower Colorado Valley, where they often form an open carpet of green, followed by unrivaled displays of flowers. On active dunes there are few ephemerals, and on the stony plains they are usually widely spaced. Many of the winter ephemerals range only to the limits of the desert or a very short distance beyond it. In the neighboring regions the spring display of verdure and flowers is almost wholly derived from herbaceous perennials.

## Summer Ephemerals

The herbaceous plants that appear after the first heavy rains of summer are as abundant and ubiquitous as the winter ephemerals. After successive rains they carpet the plains and gentle slopes in places where the surface had previously been bare for three or four months. The seedlings begin to appear about three days after the first heavy rain, and their growth is rapid. In the Arizona Upland there are very few years in which conditions are favorable for their start earlier than July 10. A few colonies of summer ephemerals have been observed after the preseasonal May rains, but these plants do not persist until the usual summer rains.

The season for summer ephemerals is longer than for winter ephemerals; the moisture of the soil is higher and more continuously maintained. The majority of plants reach a height of 12 to 16 inches (30 to 40 cm), and a few commonly grow to 6 feet (2 m). They grow more abundantly on disturbed ground than in wholly natural situations and have many of the characteristics of weeds. Grasses are much more richly represented in the summer than in the winter, both in species and in individuals. The number of species of summer ephemerals is smaller than the number of winter species, but their role in the vegetation is just as important.

Distributional features of the winter and summer ephemerals are exactly opposed; the latter are abundant in the desert areas of western Texas and diminish rapidly in western Arizona.

# Perennials

## Herbaceous Perennials

The herbaceous perennials include a wide variety of plants: grasses, grasslike plants, and broadleaf herbs. They may flower in winter or summer or both. Some like the desert lily (ajo lily) and the onion have underground bulbs, others have thickened roots, and some exhibit no storage characteristics. The desert lily is especially interesting because of its flowering in relation to rainfall. The deep-seated bulbs growing in sandy soils do not put out stems except when moisture penetrates deeply. The plant only flowers in years with ample rainfall to support growth. The undulate margined linear leaves spread out on the surface, and from them arises a stalk bearing a flower resembling an Easter lily.

The grasses belonging to the genus *Hilaria* are also of great interest. One called tobosa grass forms large communities in playa soils, especially in the Chihuahuan Desert, but also may be found growing in upland situations outside of the desert. Shreve has pointed out that, although it covers large areas in northern Mexico, it should be considered as belonging to the desert rather than to grasslands.

The largest member of the genus, big galleta grass, is noted for its ability to survive in some of the lowest precipitation sections of the Sonoran Desert and on stabilized sand dunes. Other grasses both annual and perennial appear to be at home in the desert but are nowhere dominant over extensive areas.

## Semisucculents

This group of plants is characterized by soft, usually short stems and long fibrous leaves. The leaves may be spine tipped, have lateral spines (often recurved), or both. They all produce conspicuous flowers on stalks extending beyond the leafy stems.

They are mostly found under the more favorable desert moisture conditions, and their range extends into grass and brushlands beyond the desert. They have limited water storage capabilities but do not have the mechanism for the change of volume found in the Cactaceae. For survival they depend on resistance to water loss and ability to endure drought.

**Agaves.** Agaves are also known as century plants, maguey, mescal, lechuguilla, (Fig. 7.1), and amole. The agaves, typically American in distribution, include a large number of species distributed through the warmer and usually drier sections of the Americas. Classification is difficult because of integrating species and hybridization, and therefore the total number of species is uncertain. Shreve and Wiggins list twenty-one in the Sonoran Desert—only a fraction of the total number of 250 species or more found in the United States and Mexico with the greatest number in tropical and subtropical Mexico.

The typical agave plants around Tucson, Arizona, consist of a cluster of evergreen leaves spreading out at ground level or slightly above from a very short subterranean stem (caudex). The leaves are succulent, thick with a very sharp terminal spine, and with spines or threads along the margin. Tall flower stalks are rapidly produced after several years' growth, and after flowering and fruiting the plant dies.

Agaves are used for such products as fiber, food, drink (pulque, mescal, tequila) soap, and medicine.

Fig. 7.1. Lechuguilla in foreground, ocotillo at far left, and cholla at far right.

Fig. 7.2. Banana yucca with brittlebush.

**Beargrass.** Beargrass (sometimes called sacahuista) is a grassy-appearing plant belonging to the genus *Nolina* in the lily family. Mostly the plant's short stems have many narrowly linear leaves with minute teeth on their margins. Flowers of a lily type, small and white or greenish, are produced on stems that persist long after flowering.

**Sotol.** The most common species is characterized by a short stem bearing many long narrow leaves with forward-directed prickles. The small male and female flowers are on 10- to 15-foot (3- to 5-m) stalks.

Sotol is widespread on rocky hills and bajadas at the upper edge of the desert extending into grassland and oak woodland. It has very little water storage capacity and largely depends for survival on the ability of the leaves to retard transpiration.

**Yuccas.** The yuccas (Fig. 7.2) are a large group of plants either stemless or with treelike stems such as on the soap-tree yucca and the Joshua tree. Wiggins and Shreve list nine species for the Sonoran desert, but there are many more, especially in Mexico where they are under study for commercial development.

Typically yuccas have long, narrow, fibrous, spine-tipped leaves. The white or cream-colored showy flowers are borne on stalks which, if they are within reach, may be browsed by cattle.

**Palms.** Palms are not a major element in the vegetation of North American deserts. The five species of palms in the Sonoran Desert, sometimes reaching a height of 60 feet (20 m,) grow along watercourses, around seeps or potholes in rocky canyons, and on bajadas where the water supply is abundant. Although there are no special storage organs, a considerable amount of water may be held in the trunks. The leathery quality of the leaves reduces loss of water by transpiration.

## Succulents

A major type of plant peculiar to deserts is the succulent, exemplified by the cacti. In this group of plants, the branches may be entirely lacking, the leaves may appear only as rudiments which are quickly cast off, and the entire shoot may have the form of a simple upright, cylindrical or globular body. The root systems of such plants lie immediately beneath the surface of the soil and extend horizontally over a comparatively large area. Innumerable rootlets and root hairs are formed soon after a wetting of the soil and a comparatively large quantity of water may be taken up in a short time.

When the rains cease, the finer roots die and, along with other roots, serve as anchors until the beginning of the next rainfall. The sap concentration of succulents is relatively low, whereas the actual moisture in the plant may be very high.

According to MacDougal, "The juice of one of these succulents, the barrel cactus, in the Arizona deserts is so low in mineral content that it may be used as a substitute for water by the traveller, and it is possible to obtain a quart of the cool fluid from the crushed pulp in a few minutes, as I have often demonstrated."*

The enormous accumulation of water in cacti and other succulents raises questions about the part such liquid may play

*Many other desert scientists dispute the value of this water, but the development of a simple still, consisting of a hole in the ground covered by a funnel shaped plastic sheet, has made it possible to obtain pure water from the barrel cactus.

in the life of the plant. Observations to test the matter were begun in 1908 by MacDougal and others (MacDougal and E. S. Spalding, 1910; MacDougal, Long, and Brown, 1915).

A saguaro near the Desert Laboratory with a single cylindrical trunk 18 feet (5.5 m) in height was cut down. The total weight was nearly a ton (900 kg), and a section was found to contain over 91 percent water, showing that the entire plant held over 1,700 pounds (765 kg) of water.

There are many globase to columnar cacti varying from those barely showing above ground to the tall saguaro and cardon. The barrel cactus is one of the commonest members of this group and together with the saguaro was the subject of various studies by the Desert Laboratory scientists.

Cannon (1911) noted that, although transpiration of woody desert plants increased with increased moisture supply, such is not the case for cacti unless leaves are present, a condition that occurs only during a short period in the spring. The result is that cacti hoard water, whereas the woody plants tend to expend it.

He also noted that the saguaro and the barrel cactus both have extensive shallow lateral root systems, but the saguaro also has deep penetrating roots and hence is largely restricted to deep rocky soils; the barrel cactus, however, is also found on shallow soils even when caliche is close to the surface. Root system characteristics might well explain differences in distribution.

The anchoring roots and lateral roots of the saguaro are separate systems, but large laterals are important in support as demonstrated by uprooted specimens whose laterals have become diseased.

**Saguaro.** Much has been written about the saguaro since the beginning of the century. The scientists at the Desert Laboratory nevertheless devoted great attention to this outstanding desert plant. Shreve wrote of the saguaro:

*There would have to be a chapter on its geography, on its selection of places to grow, on the nature of its root system and its method of securing water, on the fluctuations in its volume with changes of water supply, on its habits of growth, of branching and blooming, on its flowers and seeds, and on the ravaging of its seed crop by ants and birds, on the germination of the seed*

*and the vicissitudes to which the seedling is subjected. Then there would have
to be another series of chapters on the woodpecker which makes its nest in
the cool recesses of the saguaro, on the little owl which later takes possession
of the nest, and on the other birds which make their home on the saguaro
or depend on it for food in the most trying part of the year. After all has
been said about the animals to which the living saguaro is host, some more
chapters would be needed to tell about the creatures that live in and under
the fallen giants at the various stages of their decay. Then, finally there would
be much to tell about the many ways in which the Indians utilize the woody
skeletons, the fruit and the seeds, to say nothing of all the lore and traditions
that have grown up about the saguaro. In many ways it is the most important
to the aborigines of any plant native to the region. [Shreve, 1931a]*

The range of the saguaro coincides closely with the
boundary of the Sonoran Desert from the Colorado River east
and south to the valley of the Río Sonora (see Fig. 7.3). Its
altitudinal limit in this stretch is about 4,000 feet (1,200 m), but
occasional individuals are found as high as 5,000 feet (1,500 m)
on steep south slopes. In southern California the saguaro is
found in only three localities close to the Colorado River, and
it has never been observed in Baja California. In northwestern
Sonora it is absent from the sandy plains adjacent to the Gulf,
but is abundant on most of the low mountains. Heavy stands
are increasingly rare south of Tiburón Island, and it has not
been reported near the coast in the area between San Carlos
Bay and the mouth of the Río Yaqui. In the interior south of
the Río Sonora, the stands of saguaro are infrequent and open.
Southernmost individuals have been noted on hills about 9
miles (15 km) south of Cajeme (Ciudad Obregón), Sonora. It
has been reported as occurring in the lower valley of the Río
Fuerte in northern Sinaloa, but the observation is open to
doubt.

Shreve observed that the finest stands of saguaro were on
the coarse outwash slopes of mountains in southern Arizona
(see Fig. 7.4), south of the Gila River and east of Ajo, Arizona.
The size and abundance of the plants were greatest on the
southern or southwestern aspects and on lower slopes of the
adjacent hills. In no part of its range was the saguaro found on
floodplains or alluvial flats. Its best habitats were localities in
which there was rock or a soil well filled with large angular
rocks—in either of which situations firm anchorage was sup-

plied for the root system. A single large colony was seen on coarse level soil between Chandler, Arizona, and the Gila River; at this location the cactus was associated with desert saltbush. This colony was later completely destroyed by the clearing of the land for cultivation.

The habits of the saguaro—in the central part of its range at the Desert Laboratory and followed to its northern limits— merited only one explanation: it is preeminently an inhabitant of rocky slopes and southern exposures and its seeds germinate only at high temperatures. Its behavior at the limits of its northern growth, and at the limit of its growth in altitude, demonstrate that it is a plant of high temperature requirements (Spalding, 1909c).

Spalding concluded that the saguaro was essentially a subtropical species and could not, under existing climatic conditions (or any that are supposed to have preceded these conditions), have originated north of its present limits. If it has

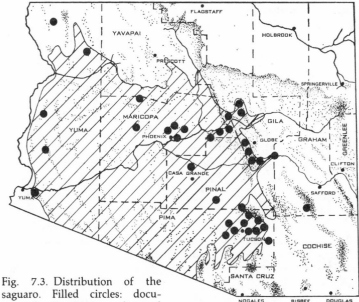

Fig. 7.3. Distribution of the saguaro. Filled circles: documented distribution; diagonal lines: distribution based on field observation. After Benson, 1969.

migrated, the general course of advance has been northward rather than southward, and it is entirely conceivable that this plant has never, within its history as a distinct species, undergone any extensive migrations.

The capacity for utilization of light rains and for persevering with an exceedingly meager annual precipitation adapts the saguaro to a fairly wide range in the southwestern United States and northern Mexico, but this range is greatly limited by its inability to cope with low temperatures. A lack of capacity to endure cold restricts its distribution locally and to higher altitudes and latitudes for, within these limits, the successful occupation of northern exposures is impossible.

The height of the saguaro at maturity is 30 to 40 feet (10 to 12 m), or slightly more in exceptional cases. The young plant is a grooved cylindrical column, rising from a slender base 6 to 8 inches (15 to 20 cm) in diameter, and reaching a thickness of 12 to 16 inches (30 to 40 cm) at a level well above the middle of the plant. Branches usually appear on plants 15 to 25 feet (5 to 8 m) high, and there are frequently five or six primary branches—although the number may be much greater in the oldest individuals. Secondary branches are found only on large plants and are never numerous. The water content is from 75 to 95 percent of the green weight, being greater at the top and less at the base. Old trees with numerous branches have been calculated to weigh as much as 6 or 7 tons (Shreve and Wiggins, 1964).

The root system of saguaro is little more than 3 feet (1 m) in depth but extends laterally near the surface for as much as 100 feet (30.5 m). A stout but elastic woody skeleton extends from base to top and consists of a cylindrical group of rods, anastomosing to middle height and free above. The remainder of the plant is a soft parenchyma of high water content, richly impregnated near the base with crystals of calcium silicate. Chlorophyll-bearing tissue blankets the plant from the top to the rough gray bark that clothes the slender base. Clusters of very stout spines are borne on the ridges of younger plants, but slender limber ones take their place at higher levels on old plants.

The superficially placed root system of the saguaro utilizes light rainfall with remarkable promptness; over certain parts of its area of distribution this rainfall probably amounts often to no more than 2 or 3 inches (5 to 7.5 cm) a year. Its storage system is perfectly adapted to the requirements of such a situation, being adjustable to either a heavy downpour that may occur in the form of torrential summer showers or to the more usual light rains that are separated by long intervals of drought.

It has been observed that throughout the life of the individual plant there is a close relation in root development to its needs for adequate support, sufficient water, and proper aeration (Spalding, 1909c). The roots are not fixed but go through gradual changes, so that what is characteristic to the youth of the plant no longer holds for the mature form. In the young plant, the main root system is that associated with the tap root that functions both for anchorage and for absorbing water. But with the growth of the plant, lateral roots arise, extending far from the parent root and taking over the function of absorbing water—leaving support of the stem to the tap root. In the larger plant, the tap root alone is not sufficient to anchor the plant against winds that at times are violent. Tap roots receive assistance from the enlarged base of the lateral roots to bear the strains and stresses set up in high aerial portions of the saguaro.

Cannon (1911) observed that when laterals die and decay before the destruction of the above-ground portion of the plant, the plant is thrown to the ground. Safe anchorage for the plant is a principal factor in limiting the distribution of saguaro, and such anchorage can be obtained only where the favorable substratum is always firm and as deep as 40 inches (100 cm), or only slightly less.

Cannon also pointed out that the roots of saguaro are generally far removed from those of other plants. Hence, intensive competition for water probably does not occur, but the plant is not entirely free from competition. The annuals found at different seasons of the year do not, as far as observed, send roots deeper in the malpais, and therefore, though their roots include a vertical range greater than that of saguaros, they compete actively with saguaro roots only in the uppermost portion of the soil.

The saguaro has a well-developed water storage system, seemingly correlated to its columnar structure which is conspicuously fluted with strong ribs and corresponding furrows that extend from base to the apex of the stem. This structure

suggests a ready adjustment of the plant by a bellows-like action of the ribs and furrows to change its bulk according to varying amounts of the stored water. Such an adjustment must be in harmony with a general mechanical system of plants that may reach a height of 50 to 60 feet (15 to 18 m) from a narrow base and are subjected to the force of high winds from every point of the compass. An examination of the stem cross section shows that there is a heavy band of thick-walled subepidermal cells both exceedingly strong and elastic. Between this band and the ring of fibro-vascular bundles, the tissue consists of thin-walled water-supporting cells.

Mrs. Spalding concluded on the basis of structure that a change in bulk corresponding to varying quantities of water contained in the tissue would manifest itself externally by expansion or contraction of the circumference—effected by folding or unfolding of the ribs and furrows. She developed some simple but effective measurements. Markings with India ink were made on points opposite each other on adjacent ribs: calipers were used to measure the distance between them at intervals of a day or more. Opposite points of the bases of the ribs were also marked in a similar manner. The thickness of the ribs between these points was also measured by calipers. Finally, a wire was placed around the plant with a coil spring partially stretched and inserted between its ends. The variation of the length in the spring indicated any change that took place in the circumference.

Mrs. Spalding found that the ribs and furrows expanded and contracted at the same time and that their action was coincident with a corresponding increase or decrease in the circumference. She concluded that the change in circumference of the stem of the saguaro is accomplished by action of the ribs, which draw closer together as the circumference decreases and move farther apart as it increases. The movements are directly correlated with the increase and decrease in the water supplied

Fig. 7.4. Typical Arizona Upland Desert view: saguaros in the background, chollas at center sides, barrel cactus in the foreground, and prickly pear in the lower left corner. The low shrub is triangle leaf bursage.

to the plant and with corresponding changes in the quantity of water held by its storage tissue (E. Spalding, 1905).

The flowers of the saguaro are borne near the apex of the trunk and larger branches. They open during the second or third week of May and are chiefly pollinated by bees. Fruits develop rapidly and are mature by the middle of June. The number of fruits on a large plant may be as great as 200, and there are approximately 1000 seeds in each fruit. Seeds are borne in a sweet, juicy pulp that is eagerly consumed by birds. In the case of seed-eating birds, this means the destruction of all the seed eaten. In the case of other birds, the seeds are passed without loss of viability. A heavy toll of the seed crop is taken by ants, as they remove the small embryo and store the seeds. The seedling is very small and provided with a short, poorly branched root, and many seedlings are covered or washed out during the rains immediately following germination. It is doubtful whether more than half a dozen seedlings persist to the following year from a crop of many thousands of seeds (Shreve and Wiggins, 1964).

The early growth rate of saguaro seedlings is very slow. After two or three years, the plants are only a fraction of an inch (a few mm) in height. The later growth is so variable that the age of a plant 3 feet (1 m) high may be from 20 to 50 years. (Table 7.2 gives typical growth rates.) After a height of 6 to 10 feet (2 to 3 m) is reached, the growth rate is about 4 inches (10

Table 7.2. Growth Rates of Saguaros

| Height | Age | Height | Age |
|---|---|---|---|
| 4 inches (10 cm) | 8.0 years | 13 feet (4 m) | 54.0 years |
| 8 inches (20 cm) | 12.5 years | 16 feet (5 m) | 60.0 years |
| 16 inches (40 cm) | 19.1 years | 18 feet (5.49 m) | 74.0 years |
| 32 inches (80 cm) | 27.3 years | 20 feet (6.10 m) | 83.0 years |
| 3.3 feet (1 m) | 30.3 years | 25 feet (7.62 m) | 107.0 years |
| 6.6 feet (2 m) | 40.5 years | 30 feet (9.14 m) | 131.0 years |
| 10 feet (3 m) | 47.5 years | 35 feet (10.67 m) | 157.0 years |

SOURCE: Shreve (1910b) and Hastings (1961).

cm) per year in the main trunk and slightly less in the large branches. Largest individuals from the known growth rate have been estimated to be from 150 to 200 years old. Wounds and infections are quickly walled off by the formation of callus. Nests made in the plant by woodpeckers are lined with a heavy bag-shaped callus, often to be found in sound condition in the remains of fallen trees. The principal natural cause of injury is the breaking of branches by high winds. Broken surfaces heal quickly in dry weather, but in a rainy period infection can set in and lead to the rapid death of the entire plant. The principal cause of death, however, is overturning by wind or by the removal of soil from the base—sometimes the softening of soil by prolonged rain.

**The barrel cactuses.** Shreve and Wiggins list twelve species and several varieties belonging to the genus *Ferocactus* in the Sonoran Desert. They are found only in regions with summer rainfall. They vary in height from less than 2 feet (60 cm) to more than 12 feet (3.6 m). The two most widely distributed are the common barrel cactus and the California barrel cactus. Most of the studies carried out by the Desert Laboratory scientists were with the former.

Like the saguaro, the barrel cactus (Fig. 7.4) is classified as a columnar cactus and, except for height growth and branching, shares with the saguaro a great many characteristics that make for success in a desert region. The most important of these are an extensive shallow root system, a large water storage capacity, and resistance to dessication.

Young seedlings are very slow growing. Shreve observed a growth of 3 inches (75 mm) in height and 2½ inches (60 mm) in diameter in four years. This rate was rapid as compared with the saguaro which had an average growth of 0.2 inch (5 mm) per year.

The barrel cactus is noted for its ability to withstand desiccation. One specimen was taken into the laboratory and studied over a six-year period. During this time it lost an average of 0.20 ounces (5 g) per day for a total of 24 pounds (11 kg). At the end of the study it was still alive but not replanted because it was thought that an anatomical and chemical examination to determine changes was more important.

**Prickly pears and chollas.** The genus *Opuntia* has the greatest number of species of any genus in the cactus family. It also shows greater variation in size, form, species, coloration, and habit of growth than any other group. One common characteristic is that each species is made up of segments or joints of approximately the same size and form. The presence of glochids (minute sharp, pointed, barbed bristles) is a feature found only in this genus.

The two major subgroups are the prickly pears and the chollas (Fig. 7.4). The first are well known for their flattened stems or pads and the latter for their more erect and branching habit and spindle-shaped joints.

The prickly pear tends to be a shrubby plant 4 to 6 feet (1.5 to 2 m) tall and 4 to 15 feet (1.5 to 5 m) in diameter. Joints or pads are 10 to 16 inches (25 to 40 cm) long and 6 to 10 inches (15 to 25 cm) broad. Small leaves are present on young joints for only a short period in the spring. The 1- to 3-inch (2.5 to 7.5 cm) long spines are dark brown usually in groups of one to three. The 2- to 3-inch (5 to 7.5 cm) diameter flowers are commonly yellow with a green center composed of stigmas. The fruit is fleshy 1 to 2 inches (2.5 to 5 cm) in diameter and 1 to 3 inches (2.5 to 7.5 cm) long. It becomes red or purple at maturity and is prized for jelly and wine making.

The principal adaptive characteristics of the prickly pear are its shallow spreading root system and its water storage capacity. This combination makes it possible for the prickly pear to take up water rapidly after rains, even though the soil is moistened to a depth of only a few inches and to store it for future needs. Also the combination of nighttime accumulation of carbon dioxide needed for photosynthesis and the closure of stomata during daylight hours naturally reduces water loss.

Prickly pears have the greatest geographical range. They inhabit a wide variety of localities and soils and withstand all but the most arid situations. Smaller members of the group range northward into Canada and eastward to the Atlantic coast while the larger members are found in nearly all plant communities in the warmer climates. Shreve and Wiggins list seventeen species in the Sonoran Desert. Because of the difficulty of classification, other authors may list more or fewer.

The Engelmann prickly pear, common throughout most of the Sonoran Desert and ranging into Nevada and Utah and the Chihuahuan Desert, received the most attention by scientists at the Desert Laboratory.

The chollas include a wide variety of forms (Fig. 7.4). The only anatomical feature they have in common is the round joints which may be fleshy or not and spiny to almost spineless. They vary in height from 1 to 15 feet (.3 to 5 m) and usually are much branched. The spines have easily detached sheaths; glochids are small and inconspicuous. As described earlier, chollas are notable for the variation of flower color and, for some, the time of day for flowers to appear. There are twenty-one species in the United States and thirty in the Sonoran Desert.

The staghorn cholla received the most attention at the Desert Laboratory. It is a shrubby openly branched plant 3 to 12 feet (1 to 4 m) tall with a woody trunk from 2 to 5 feet (.6 to 1.6 m) feet high. The slender cylindrical joints are green or tinged with purple. The flowers, 1 to 2 inches (2.5 to 5 cm) in diameter, are yellow, orange, greenish red, brown, or bronze.

Mrs. Edith Shreve (1915–16) was particularly interested in its water relations and the "activity" of its joints. The movement of branches which may occur daily was related to moisture content—the branches becoming more erect as pressure from contained water increased and drooping as internal pressure decreased.

Cannon (1915) observed that the best conditions for root growth were the warm temperatures and good aeration of the upper soil layers. On the basis of wide observations, he concluded that staghorn cholla and other chollas found conditions most favorable to their growth under a combination of summer rains and moderate to light textured soils.

He also noted that the pencil cholla had the most dense network of roots in all of his observations. So dense in fact that it occupied the soil area to the exclusion of other plant roots.

## Woody Plants

The woody types of plants may be seen away from streamways. They are species of annuals and perennials with hardened stems; branches reduced to spines; and small, narrow, hardened, and waterproofed leaves that extend their roots in a kettle-shaped mass to only a moderate depth in the soil.

Desert soil moisture available to plants is in the more superficial layers that are wetted by the rains. The spinose plants contain a very small proportion of water; their bodies are hard, with a minimum of development of cortex or pith. Although they hold only a small amount of sap, the juice generally contains a large proportion of salts or compounds that exert osmotic pressure (MacDougal, 1911i).

MacDougal noted that plants of this type are constantly in absorbant contact with the soil, and apparently continue to derive some water from it even in the driest times—as evidenced by the fact that they wilt quickly when taken up.

Almost any branching plant with broad leaves will, if forced to carry out its development under arid conditions, show some features of the type of desert plants described. MacDougal believed that it was perfectly safe to assume that spinose forms represented the simplest or most elementary specializations of desert plants, and species with the most diverse morphological constitution might show alterations of this character. He listed examples of these as: mesquite, catclaw, fairy duster, paloverde, ironwood, creosote bush, gray thorn, and ocotillo, among others.

**Bursages.**    Two bursages, the white bursage and the triangle leaf bursage, are important components of the vegetation of warm deserts. The white bursage may be the most common plant in the western portion of the Sonoran Desert and southeastern portion of the Mojave Desert, or with creosote bush make up 95 percent of the perennial vegetation.

The hemispherical shaped plants have small deeply divided leaves often so dry it is difficult to tell if they are alive. In late spring many of the leaves die but may remain attached to the plant. The leaves quickly renew growth after rainfall. Under favorable conditions the growth of seedlings is rapid with roots quickly formed 5 to 15 times the length of the stem.

Triangle leaf bursage (see Figs. 7.5, 7.6) is not as widely distributed but was more common in the vicinity of Tucson and hence received more attention by the Desert Laboratory workers. It occurs only on rocky soils having a covering of coarse rocky fragments, whereas white bursage is found on sand or fine-textured soils. This low-growing shrub has a compact crown with either relatively large or small leaves depend-

Fig. 7.5. Triangle leaf bursage often occurs in pure stands.

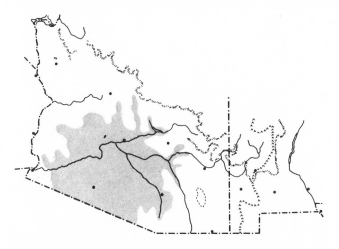

Fig. 7.6. Distribution of bursage. After Benson and Darrow, 1981.

ing upon moisture conditions. During the rainy season the large leaves produced in abundance turn gray with age.

Bursage lateral roots develop short rootlets 1/2 to 2 inches (1.75 to 5 cm) in length following rains. These enable the plant to absorb moisture rapidly from moist soil. They cease activity when the soil becomes dry.

**Brittlebush.** Brittlebush, one of the most widely distributed plants in the Sonoran Desert, extends into the Mojave Desert and the thorn forest of Mexico (Fig. 7.7). It is most abundant, however, in the northern half of the Sonoran Desert. In Arizona it is characteristic of coarse outwash slopes and bajadas; farther south it is found only on the sandy soils of alluvial bottoms. In Baja California it frequently occurs in pure stands.

Brittlebush endures the hottest temperatures in the Sonoran Desert, but does not reach the northernmost or highest areas. At Tucson, Arizona, it may freeze to the ground but generally recovers.

Brittlebush, with its large gray-green leaves and bright yellow flower heads, is one of the most conspicuous plants in the Sonoran Desert (Fig. 7.2). It may reach a height of 3 feet

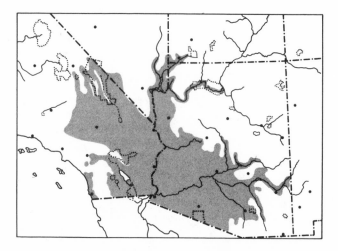

Fig. 7.7. Distribution of brittlebush. After Benson and Darrow, 1981.

(1 m) but is usually smaller. It has many brittle branches (hence the common name) that exude gum in the warm and dry portion of the year. Leaves are borne during both winter and summer rains, but may be inconspicuous or lacking in the dry season. The egg shape or triangular winter leaves are 1 to 2 inches (2 to 5 cm) long and 3/4 to 1 inch broad (1.5 to 2.5 cm). These leaves are usually shed during late spring and replaced by much smaller leaves covered with a thick mat of hairs.

The bright yellow flowers are produced on elongated leafless stalks in such abundance as to give hillsides a yellow appearance. Many seeds are produced and give rise to many seedlings, most of which do not survive to reach maturity; thus, the population, although made of mixed ages, never exceeds the environmental capacity.

Mrs. Edith Shreve became very interested in brittlebush and its ability to reduce transpiration during hot dry periods. Part of this was accounted for by the reduction of leaf surface, but it also appeared there was less transpiration as the water content of the leaves was reduced. Thus, as soil moisture became less, loss of water from the plant was reduced.

In times of drought the leaves and stems are filled with a dark brown viscous sap which oozes from every wound during dry periods but is absent during the moist season.

Brittlebrush has a generalized root system which enables it to adjust to the various soil and moisture conditions of the Sonoran Desert. This characteristic, together with its variable foliage, is important to its widespread distribution.

**Jojoba.**    Jojoba, an evergreen 3 to 15 feet (1 to 5 m) tall desert shrub, has come into prominence, because it is the only known plant that produces wax rather than oil. The wax is almost identical with that found in the sperm whale, an endangered species. Because of potential values for a wide variety of products from cosmetics to high-speed high-temperature lubricants, it is being developed commercially.

This evergreen shrub with leathery leaves is found on outwash slopes and arroyos from southwestern San Diego County in California to the eastern edge of the Sonoran Desert and southward into Baja California and Sonora. It is drought tolerant but subject to frost damage in colder winters.

Fig. 7.8. Creosote bush.

**Creosote bush.** Creosote bush (Fig. 7.8) is one of the most abundant and widespread shrubs of the Sonoran Desert and is equally prominent in the vegetation of the Chihuahuan Desert, the Mojave Desert, and a small part of the Great Basin Desert (Fig. 7.9). Growing from elevations below sea level in Death Valley to 7,955 feet (2,625 m) in the mountains of Zacatecas, it persists in regions where twelve months frequently pass without precipitation. Irrespective of differences in the seasonal distribution of rainfall, creosote bush occurs in all sections of the desert and even invades the edges of grassland and encinal areas where annual precipitation can reach 20 inches (500 mm). It grows on well-drained deep alluvial soil, on the edges of alkaline flats, on sandy plains, and on the rocky slopes of volcanic hills. The creosote bush's adaptability to a wide range of conditions is greater than that of any other desert plant. Height, density of branching, size of leaves, and rate of growth change substantially under different conditions, since the plant responds readily to an increased water supply. Near the edges of its range, the bush is restricted to well-marked habitats, but throughout the center of the desert it is less restricted and, even in the most unfavorable areas, is found scattered everywhere.

The northern distribution limit of creosote bush is in southern Nevada and southwestern Utah, where it follows approximately the contour of 4,000 feet (1,200 m). According to Shreve (1940) the northern range of creosote bush is near the isoclimatic line for a maximum of six consecutive days of freezing temperature. Heavy snowfall is distinctly hostile to creosote bush. The configurations of the plant—the inclination of the branches—are such that the weight of snow on the twigs and leaves flattens the entire plant and it never completely recovers its former position. Northernmost plants in New Mexico have been so flattened by snowfall that they suggest creeping perennials.

In Baja California, creosote bush is confined to the eastern side of the northern mountains but extends across the peninsula to the ocean near Rosario. It extends south to the vicinity of La Paz and Todos Santos but is not found on the southern mountains. At its eastern limit in Sonora and in the belt that connects the ranges in the Sonoran and Chihuahuan deserts, creosote bush occurs in large and small isolated colonies.

Creosote bush reaches its southern limits in North America on the plateau of Mexico in regions with a fairly dependable annual precipitation of 12 inches (30 cm). It does not, however, pass beyond the edge of the desert in Texas and does not reach the edge of the desert in southern Sonora. Shreve found the southernmost known colony of creosote bush near the coast about 20 miles (32 km) southeast of Guaymas. On similar terrain, and with a very similar climate, it extended 270 miles (434 km) farther south in Baja California, where it was finally limited just north of Todos Santos by the prevailing granitic soils of the Cape Region.

Well-established seedlings, free from competition, will grow in a soil of constantly high moisture content at a rate that is roughly seventy times the rate under adverse soil conditions. However, no part of North America is too dry for creosote bush; it persists in extensive colonies in Baja California, for example, where there have been periods of four years without enough precipitation to wet the soil to a depth of 0.4 inch (1 cm).

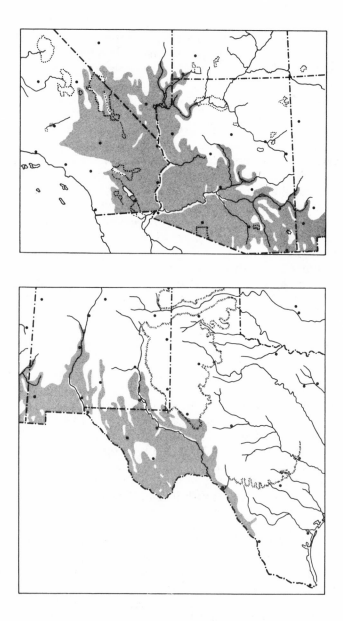

Fig. 7.9. Distribution of creosote. After Benson and Darrow, 1981.

The highest summer temperatures where creosote bush grows occur in Death Valley and the Salton Sink. Since these are the warmest localities in North America, it is obvious that high temperatures do not limit its native distribution. The rapid disappearance of creosote bush at the northern and western edges of the arid bushland of Mexico indicates that here conditions are not suitable for germination and establishment, and it is unable to persist in competition with vigorous shrubs of greater stature. It has been seen at an elevation of 8,600 feet (2,620 m) on the Picacho de las Bocas, in Zacatecas, where the rainfall is probably 20 inches (50 cm), but it occurs there only on thin stony soil where it is both unshaded and without competition.

One of the earliest observations on creosote bush was made by Coville during his 1893 studies in Death Valley. He was impressed by the appearance of creosote bush, which has

*the apparently simplest form of resinous coating. The leaves and small twigs are thinly spread with a covering that closely resembles in appearance ordinary shellac. To the abundance of this resinous material the plant's popular name creosote bush is due, for in burning the green wood and leaves of Larrea, a pungent odor is detected and dense smoke arises. That the function of the coating is to minimize transpiration there can be no doubt, but the precise method by which this is brought about has not been ascertained. If it were simply by the complete mechanical varnishing of the leaf surfacing, all transpiration would cease. It should be pointed out here that in winter, when we first became familiar with the creosote bush, its leaves were thoroughly varnished; but in June when the spring growth had nearly ceased, the leaves appeared to have very little of the coating. There is in fact an evident correlation between rapid transpiration and absence of resinous covering and a similar correlation between slow transpiration in the presence of such covering.*

Spalding (1904) was impressed with the ability of the creosote bush to accommodate to available water supplies. When individual plants were well watered, the leaves became a deep green and underwent a marked increase in size, contrasting greatly with plants around them that did not receive water. While the well-watered plants presented the appearance of robust health and vigor, the others appeared pinched, with narrow, pale leaves, more or less defoliated branches and other marks of a serious struggle for survival. He interpreted this as an indication that the creosote bush—during the course of its development under greater precipitation—had acquired the ability to withstand excessive drought but had never lost the capacity to absorb and use large quantities of water and, in fact, to attain its best development under such conditions.

Spalding made investigations of the root system of the creosote bush plants and found that the general plan of the root system included a strong tap root which grows downward until it changes its course either because it meets an obstruction or for some other reason. Additionally, he found slender lateral roots running near the surface for some distance. He believed that the root system, spreading widely and penetrating deeply into the earth, is a system well adapted for securing available moisture through a comparatively wide area when there is light rain, and for reaching moisture at lower levels after heavy rain.

The internal structure of the creosote leaves gives but little clue to the understanding of the remarkable resistance of the bush to drought. The leaf could hardly be described as xeromorphic. The cuticle is not as thick as might be expected on a desert plant; and stomata, far from being reduced or specially protected are decidedly numerous and mesophytic, being present in both the upper and lower epidermis (Runyon, 1934).

Although the creosote bush is always in leaf, there are conspicuous seasonal changes. Obvious growth is initiated only after a period of rain. If the water supply is abundant, growth is very rapid; first leaves, then flowers and fruits, are produced in great abundance. With less moisture the growth is slower and there are few flowers and fruits. Although confined in its normal distribution to the most arid parts of the continent, the creosote bush attains its maximum of vegetative vigor when supplied with abundant water.

Runyon observed that in times of drought the older leaves generally yellow and fall a few at a time and those remaining are brownish rather than green. In the very driest seasons only buds and immature leaves remain. In such cases many twigs are also shed, and, under the most extreme conditions, large older branches die, leaving fewer and younger parts to compete for the meager water supply. Whatever mature leaves remain on the branches during severe drought become yellow and fall soon afterwards. Immature leaves and buds are retained through the driest seasons in a partially dormant state, resuming their activity when conditions are again favorable.

Low moisture in the upper layers of soil may play an important role in bringing about scanty reproduction of the creosote bush. A typical creosote area is not densely populated by creosote bush, and usually the surface of the soil is relatively free of vegetation. The first 12 inches (30 cm) of soil is quite well permeated by roots, and there is no doubt some competition for moisture. However, there cannot be much where the moisture content is below the wilting percentage nearly all the time (Mallery, 1935).

The greatest number of young creosote bushes per given unit area are almost invariably found where the soil has been disturbed—along roadways and around excavations where scrapers and other implements have loosened and piled up the soil. Mallery concluded that seed germination and seedling development are aided or made possible by covering the seeds to prevent rapid desiccation and by loosening the soil to facilitate the penetration of roots. Moisture also penetrates more readily into soil that has been disturbed and there is less lost as runoff. The rain that falls is therefore more effective in maintaining the moisture content of the soil.

The creosote bush, frequently associated with the saguaro, lacks the saguaro's water-storing capacities but nevertheless is far more successful as a desert species. It can maintain itself with an exceedingly meager percentage of soil water (shown by its occupation of the creosote bush slopes) but luxuriates in an abundant moisture supply, as is clearly seen by the vigorous growth in washes and wherever there is plenty of water. The bush successfully withstands a far wider range of temperature than does the saguaro and, with these physiological characteristics, is found ranging far beyond the latter in both latitude and altitude. Its success as a desert species is a striking illustration of the great advantage of physiological endurance over highly developed adaptations in adverse circumstances (Spalding, 1909c).

One point at which even the creosote bush is at a disadvantage is indicated by the sharp line between the Tumamoc slope where creosote bush is abundant and the floodplain of the Santa Cruz where it is absent. This avoidance of the floodplain—seen in all the great river valleys of southern Arizona —of salt spots, and of some bolson areas seems to merit one explanation; the creosote bush is preeminently a plant of well-drained ground. Whatever else it can endure, it is unable to exist where there is defective aeration of the soil. Thus, this most successful plant of the desert—the one that seems adapted to a wider range of conditions than any other—is brought to a full stop at the line between its own special habitat and that, for example, of the mesquite and the saltbushes.

**Ocotillo.**  Covering an area from northern Baja California to Sonora, Chihuahua, and Coahuila, Mexico, and from western Texas to southern California in the United States, the range of ocotillo slightly exceeds the limit of the Sonoran and Chihuahuan deserts (Figs. 7.10, 7.11). Small colonies or single plants occur in mountains which overlook the northern edge of the desert.

In the Santa Catalina Mountains ocotillo occurs at 6,000 feet (1,800 m) and throughout the desert-grassland and encinal of southeastern Arizona on south exposures between 5,000 and 6,000 feet (1,500 and 1,800 m). In the Swisshelm Mountains in southeastern Arizona it reaches the highest recorded altitude for the United States at 7,000 feet (2,100 m). These higher occurrences carry the continuity of distribution from the Sonoran to the Chihuahuan deserts (Shreve, 1951; Shreve and Wiggins, 1964).

Shreve found ocotillo abundant in Baja California at desert levels on the Gulf coast as far as latitude 28°, and on the Pacific coast from the desert boundary near Rosario southward nearly to the Vizcaino Plain. In Sonora ocotillo was abundant north of latitude 29°, extending from the shores of the Gulf to the hills bordering the valley of Río Montezuma. There were many isolated occurrences of the plant far above the desert level. South of Río Sonora ocotillo was infrequent, and it was not observed south of the vicinity of Torres.

Fig. 7.10. Typical stand of ocotillo.

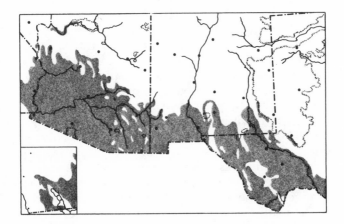

Fig. 7.11. Distribution of ocotillo. After Benson and Darrow, 1981.

Throughout most of its range ocotillo is most abundant on the shallow soil of rocky slopes or coarse outwash slopes. On the broad plains of the Lower Colorado Valley it occurs infrequently in the nearly pure stands of creosote bush. When colonies occur in the creosote bush plains, they are invariably found on unusually rocky soil, derived from small volcanic emergences. In all parts of the desert with more than 4 inches (100 mm) of rain, ocotillo is absent from fine-textured soils and subject to sheet flooding. In the drier parts of the desert, however, it is found frequently on finer soils and occasionally on sand.

Heaviest stands of ocotillo are usually found above 3,300 feet (1,000 m) along the upper edge of the desert and in adjacent hills. These stands occupy slopes gently inclined to the south and southwest having a coarse soil filled and covered with rock fragments. Individual plants grow from 8 to 15 feet (2½ to 4½ m) high. Each one can have from ten to fifteen stems, and the stem surface often indicates many young plants and active growth. In many such stands plants are so closely spaced that it is almost impossible to walk freely between them.

Generally, the root system of ocotillo is shallow and wide-spreading, much like that of the cacti, and optimum growing conditions require quick wetting and penetration in rainstorms, combined with good soil aeration at all times. The larger plants of ocotillo, as well as other shurbs, have a root system well enough distributed through the soil to correspond to their comparatively large transpiration surfaces. These species probably root very deeply in rock crevices. Seedling ocotillo roots elongate directly downward rapidly enough to reach a permanent and adequate water supply before the soil, wet thoroughly by the frequent showers of the rainy season, dries out enough to result in drought injury (Livingston, 1906c).

In 1943, Darrow, a faculty member of the University of Arizona, carried on extensive studies on the Desert Laboratory grounds on the life history and growth forms of ocotillo. He concluded that the unique appearance of mature ocotillo plants is due largely to the manner and sequence of branch development. The mature plants consisted of forty to seventy-five or more straight slender branches, 8 to 15 feet (2½ to 4½ m) in height which fanned out near the ground level from a short main trunk. The branches had few laterals and produced clusters of four to twelve leaves. The successive development of basal lateral branches by the parent branches tended to produce a fan-shaped cluster of long straight stems of various lengths radiating out from the short basal trunk. Darrow observed that in response to the two rainy seasons in the Tucson area, ocotillos come into leaf twice during the year, first with the beginning of the winter-spring rains and later at the start of the summer rainy period. Plants retain their leaves as long as the amount and frequency of rainfall is sufficient to maintain available soil moisture. In years when the rains are widely scattered, several periods of foliation and defoliation may occur. Terminal growth in the ocotillo occurs principally during the summer rainy periods. Its florescence is associated with the spring rainy season and continues into the presummer drought period in April. Darrow thought that ocotillos, which bear foliage only when soil moisture is available, are better able to accumulate food reserves during the prolonged spring rainy season than during the summer, when soil moisture might be available only at irregular intervals.

Cannon made a study of the transpiration in ocotillo during 1904. It was very dry that year—with precipitation only about 75 percent of average—the conditions were very severe but in August were favorable to plant growth. The usual winter and spring annuals did not appear and the other plant forms bore no leaves at all, or if they did, they were dropped when drought returned. In summer, however, all vegetation was very luxuriant. Even during the period of leaflessness, ocotillo maintained a feeble rate of transpiration that varied directly with the water supply. The rate increased immediately in response to the rain and decreased as the dry period following rains became greater. There was an increase in transpiring surface, but a decreased rate occurred without any immediate and corresponding increase in the transpiring surface. The least rate of transpiration when leaves were present was observed in the dry and cool period of March. The highest rate was observed August 26, near the close of the summer rains, when the temperature was high (Cannon, 1905b).

Cannon was very much surprised at the promptness with which ocotillo was able to develop leaves when the water supply of the plant was increased by rain. For example, he cited a case where several weeks previous to May 11 one of the plants studied had no leaves. Within 48 hours after the rain, it was well covered. He also applied water in a very dry period to one specimen and leaves were observed two days later; on the third day they were about ½ inch (1.3 cm) long, and on the fifth day they were about mature size (1 to 1-½ inches, 2.5 to 4 cm). One of these ocotillos in the experiment formed five distinct leaf growths between February and August. Very soon after the rains ceased the leaves were always dropped and excessive transpiration was thus avoided. At least part of the explanation for response to precipitation is that the ocotillo bark can absorb liquid water and yield it to growing tissues (Lloyd 1905a).

Among the unusual characteristics of the ocotillo observed by early workers at the Desert Laboratory was the formation of the spines. It was thought that the spines or thorns were modified petioles, or leaf stems (Humphrey 1931). These thorns are formed only by leaves that occur on elongating branches. These leaves are almost never seen except during the latter part of the summer rains. As the branches elongate, the leaves are composed of a small blade on what appears to be a very thick petiole. With the advance of dry weather and consequent cessation of growth, the apparent petiole seems to split for its whole length in a horizontal plane, and the upper one-quarter with the leaf blade attached falls off, leaving a stout, sharp-pointed spine.

Humphrey concluded that the thorn is an outgrowth of the cortex and epidermis. The petiole is attached to the spine and formation of the abscission layer, preceding the dropping of leaves, is along this juncture. On this basis the spine and the petiole should be considered as separate entities.

In 1935, Humphrey stated that the origin of the family to which ocotillo belonged was unknown; like other desert flora the fossil records are lacking. There has been some question as to whether it belonged with the extremely xeromorphic shrubs or the succulents. The ability to cut down on the transpiring surface by dropping leaves, and a heavy waterproofing of the stem surface, are more or less characteristic of xerophytic shrubs. The essential difference between the shrubs and the succulents is the presence of water storage tissue in one and the absence in the other. The members of the ocotillo family are difficult to place in either category.

**Enlarged trunk species.** There are several trees with "swollen" trunks in the Sonoran Desert. The most notable of these are the torote, boojum tree, and the copal. The first is entirely restricted to Baja California and islands of the Gulf of California. The boojum has the same distribution except for a small colony on the coast of Sonora near Puerto Libertad (Figs. 7.12, 7.13). All three are adapted to desert conditions by their extensive root systems and water storage capacity in their trunks.

The copal is much more widely distributed through the warmer portions of the Sonoran Desert. It is a handsome tree reaching a height of 30 to 35 feet (10 to 12 m). Only the lower trunk and branches are thickened. The dark green leaves tend to be evergreen and together with the smooth gray to bronze-red bark make a very attractive combination.

In some places torote is a spectacular tree with dark green leaves, a smooth buff colored exfoliating bark, but near the Pacific coast it is usually prostrate.

The boojum tree or cirio is the most bizarre plant of the Sonoran Desert. Closely related to the ocotillo it has the same type of branching and general leaf characteristics, but the branches are produced at the top of a grotesque upside down carrot shape trunk that may be 20 to 30 feet (6.6 to 10 m) in height. The trunk at its widest point may be 20 to 30 inches (50 to 70 cm) in diameter and tapering to 3 to 5 inches (7.5 to 12.5 cm) near the top. Smooth gray bark covers the trunk.

Slender thorny branches grow horizontally from the trunk in young trees, but the tall ones are usually naked. There is a single tapering trunk that occasionally divides into two or more branches. These grow upright and close together as if there were no room for them to separate. Seen at a distance the great colonies of boojum look for all the world like a burned forest in the northern Rockies. Leaves of the boojum are borne on the lateral branches or in tufts on the trunk and remain as long as the soil is moist. In the trunk there is a thick cortex, a woody cylinder, and a large central core of pith and cortex with a very light system of vascular tissue. When the soil is moist

Fig. 7.12. Boojum trees near Puerto Libertad, Sonora, Mexico.

the cells of the pith become turgid with water, but after a year or two of dry weather great air spaces may be found in the pith.

The common name *boojum* was first applied by Godfrey Sykes of The Desert Laboratory. Sykes had read Lewis Carroll's *The Hunting of the Snark,* which referred to a mythical thing called a *boojum* in desolate far-off regions. When Sykes first saw the plant he is reported to have said, "Ho, ho, a boojum, definitely a boojum!" and the name has persisted.

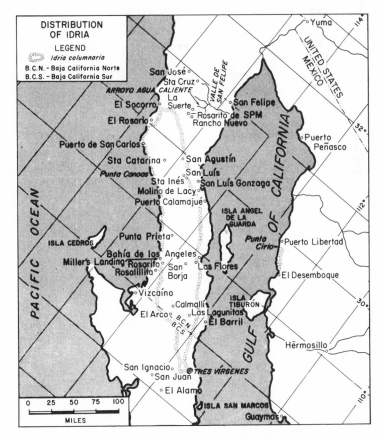

Fig. 7.13. Distribution of the boojum. After Humphrey, 1974.

Fig. 7.14. The trees on the left are foothill paloverdes.
Cacti are saguaros and organ-pipe.

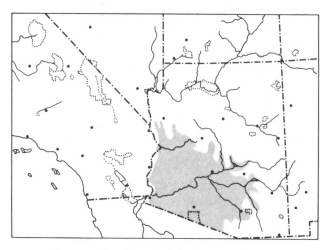

Fig. 7.15. Distribution of the foothill paloverde.
After Benson and Darrow, 1981.

**Paloverde.** There are six species of paloverde belonging to the genus *Cercidium* and one in *Parkinsonia.* The chief characteristic of these trees is that all the smaller branches and most of the larger ones remain green and are capable of carrying on photosynthesis. The foothill and the blue paloverdes received the most attention from laboratory scientists, although all are mentioned in the descriptions of the vegetation of the Sonoran Desert.

The foothill paloverde is widely distributed in the Sonoran Desert; in California it is restricted to a few localities near the Colorado River; and in Baja California it is absent from the Pacific side (Figs. 7.14, 7.15).

The foothill paloverde commonly reaches a height of 15 to 20 feet (5 to 6.6 m) and occasionally 25 to 30 feet (8 to 10 m). The trunk is short, and it branches close to the ground. It has a smooth gray bark; above the trunk the green bark may have dark patches covering wounds.

The compound leaves with three to five very small leaflets are produced after rains. Usually there are two sets but they may not appear if rains are wanting. The leaves persist for six to ten weeks, then turn yellow and are shed. It is probable that the total leaf surface is less than the area of the green stem surface.

Flowers appear in April or May provided rainfall has been average or above. In dry years the trees may not bloom. The long constricted pods ripen about six weeks after flowering, releasing one to three seeds. A large part of the seed is consumed by rodents. The seeds have a waxy coating that resists germination until the following year; if moisture and temperature conditions are favorable, abundant seedlings may appear within three days after rain. Seedlings quickly reach a height of 1 inch (2.54 cm) or more with roots two or three times as long. During the dry aftersummer period, the mortality is very high and continues to be high during the first four or five years.

In exceptionally dry years, twigs, small branches, and occasionally large limbs will be lost, but it is rare for an entire tree to die from drought. Heavily injured trees produce new shoots from the base following loss of major branches. Shreve estimated that trees live to be 300 to 400 years old.

The root system of foothill paloverde is wide-spreading, but not near the soil surface. On rocky slopes the roots penetrate narrow fissures in the rock where small amounts of water percolate.

The blue paloverde develops a 3-to 6-foot (1 to 2 m) trunk with light gray bark, and a rounded crown consisting of stout branches with scattered thorns. The leaflets are about twice the size of those of the foothill paloverde, and the foliage and branches are blue-green. The flowers are deep yellow and appear ten days to two weeks earlier than those of the foothill paloverde.

Blue paloverde has much the same distribution as foothill paloverde but extends beyond it and to a higher elevation. Blue paloverde is most abundant on fine soil, on alluvial plains, and in washes or near the bed of streams (Figs. 7.16, 7.17).

The oldest trees are 20 to 30 feet (6.6 to 10 m) tall with a wide-spreading crown. Blue paloverde reaches its maximum growth along streamways in southwestern Arizona and northwestern Sonora. Seed production and germination is similar to the foothill paloverde. In young trees the nearly horizontal branches have very thorny twigs. Its maximum age is not known but probably is not much different from that of the foothill paloverde.

The blue paloverde has a generalized root system with roots penetrating 6 feet (2 m) or more and often extending up and down the streamway 30 feet (10 m) or more.

**Cottonwood.**    The cottonwood along with other trees that derive their water supply from groundwater and hence are more or less independent of rainfall may be said to live in a desert region but not in a desert environment. However, the cottonwood is subject to high evaporation rates.

**Ironwood.**    Ironwood (Fig. 7.18), sometimes called *tesota,* is notable for several characteristics. Its distribution corresponds closely to the outlines of the Sonoran Desert. It is most common in areas with less than 8 inches (200 mm) of precipitation, but under such climatic conditions it is restricted to washes where it receives the benefit of moisture derived from runoff from higher slopes. It is strictly limited by low winter temperatures and has been considered as an indicator of suitable climate for citrus trees.

Fig. 7.16. The blue paloverde grown as an ornamental shade tree.

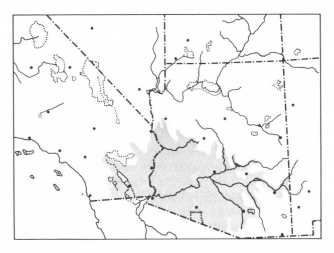

Fig. 7.17. Distribution of the blue paloverde.
After Benson and Darrow, 1981.

Fig. 7.18. Cultivated ironwood trees.

It resembles the mesquite tree, with a foliage of many leaflets, but is larger—in fact, the largest of all nonsucculent plants in the Sonoran Desert. The evergreen foliage is copious, resulting in more shade than is usual for most desert trees.

The adaptations for desert survival are much the same as those for mesquite, except that it is not found on floodplains but is restricted to bajadas and upper slopes where it is not subject to temperature inversions.

Blue pealike flowers are produced in possibly two out of five years. The pods with one or two seeds mature in June and fall shortly after maturity. The seeds germinate readily which again is in contrast with legume trees notable for their hard seed covering that inhibits germination.

The heartwood is very fine-grained and hard and has become the basis for a figurine carving industry.

**Mesquite.** Forty-four species of mesquite have been recognized. Forty of these are native to the Americas; the nine species native to North America include four in the Sonoran Desert. Only two are widely distributed in the arid portions of North America, the velvet mesquite and the honey mesquite.

The velvet mesquite is largely restricted to Arizona and northern Mexico in the Sonoran Desert, whereas the honey mesquite (in its eastern or western subspecies) has a range

Fig. 7.19. Large mesquite growing on floodplain with underground water supply.

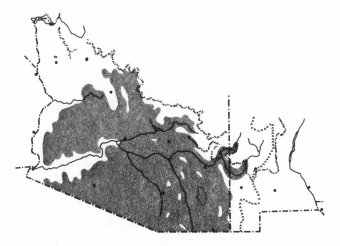

Fig. 7.20. Distribution of the mesquite. After Benson and Darrow, 1981.

including the eastern edge of California and all of Baja California to eastern Texas, northward into Kansas (Figs. 7.19, 7.20). The honey mesquite is a large shrub or small tree. Where the soil is sandy, sand blown by the wind often collects around the plants and forms small dunes, leaving only the tips of branches exposed. These buried trunks are responsible for the descriptive reference in New Mexico to natives who "dig for wood and climb for water."

The velvet mesquite is usually a tree, which may attain a height of 50 feet (16.6 m). Those that have survived the inroads of settlement, however, are usually less than 30 feet (10 m) in height and often appear shrublike with a height of 15 feet (5 m) or less.

The initiation of leaf growth in early or late March appears to be governed by temperature, and hence it has been cited as an indication of the beginning of the frost-free season. At first the leaves appear to be slightly fleshy, but by mid-April they take on a hard shiny appearance which is retained until leaf fall in October or later. In warmer portions of the desert they may remain throughout most of the year.

The mesquite flowers appear as yellow catkins. Flowering may begin at the same time that leaves appear and continue for about a month with pods appearing in May. If moisture conditions are favorable, trees may flower again after the summer rains. Mature pods may appear in June, but most of the fruit matures in late July and drops from the trees in September. Occasionally trees produce up to three bean crops during the year. A tree 10 feet (3.6 m) tall may produce 40 to 50 pounds of fruit, but usually not more than 10 to 15 pounds.

Unlike most legumes, the fleshy pods of the mesquite do not open up. The pods are especially rich in sugars and the beans in protein. They are eaten by wild and domestic animals and were prized by southwestern natives. More than 75 percent of the seeds are not digested by cattle, and the undigested seeds have a high percentage of germination and survival resulting in the rapid spread of mesquite to the detriment of rangelands, because the native grasses lose out in the competition with mesquite.

The velvet mesquite has an extensive root system which may largely be in the surface foot of soil or may extend to great depths to reach subsurface water tables. Because of the latter adaptation it has been often listed as a phreatophyte and an indicator of underground water supplies.

In contrast with many plants having only shallow root systems, mesquite roots are not dependent on warm soils for root growth and hence can penetrate to the cooler deep soils.

The velvet mesquite exhibits great plasticity which enables it to grow in a variety of habitats within the arid region. In upland areas it may depend entirely on moisture available in the upper soil layers but in lowland areas it may tap accumulated water at depths of a few to many feet.

The depth to which mesquite roots will penetrate is conditioned by the character of the soil, the depth of the water table, and, especially in upland areas, by the penetration of rains. The variation in size of mesquite trees has been thought to be related to the depth of the water table: where the mesquite is large, the perennial water supply is relatively close to surface and, where trees are small, the water supply is deep or limited.

Even though mesquite is notable for its deep tap root, its system is actually of the generalized type with both tap and lateral roots well developed. It is the development of lateral roots that makes possible its spread into upland areas.

The growth of mesquite roots is not seriously reduced by poor aeration. The capacity of mesquite roots to reach great depths is explained by this characteristic, which contrasts with plants like cactus which have shallow root systems and which require the presence of oxygen for root growth. Compared with cactus roots, mesquite roots also have a relatively rapid growth rate at low temperatures.

The early Spanish explorers mentioned mesquites growing along main drainage, and it was not until 1890 that they were reported to be invading adjacent desert slopes and even into grasslands as far north as Albuquerque, New Mexico.

The larger trees that were abundant in small forestlike bosques provided building materials. More recently the principal products have been fence posts and fuel. The dark hard wood is often used for carvings, and mesquite ranks next to ironwood as a choice wood for cooking and heating.

# Plant Names

*These lists of common names–scientific names and scientific names–common names are intended only for reference in this book. Neither scientific nor common names are authoritative. Scientific names are those used in Desert Laboratory publications, and common names were chosen on the basis of common usage. Scientific names for the Sonoran Desert follow Shreve and Wiggins (1964).*

## *Common Names–Scientific Names*

| | |
|---|---|
| acanthambrosia | *Franseria* |
| acarospora | *Acarospora* |
| alder | *Alnus* |
| alfileria | *Erodium cicutarium* |
| alligator juniper | *Juniperus deppeana* |
| amaranth | *Amaranthus* |
| ambrosia bursage | *Franseria ambrosioides* |
| American mangrove | *Rhizophora mangle* |
| American trixis | *Trixis californica* |
| American vetch | *Vicia americana* |
| amole | *Agave schottii* |
| anacahuita | *Cordia boissieri* |
| Anderson lycium | *Lycium andersonii* |
| annual haplopappus | *Haplopappus gracilis* |
| annual lupine | *Lupinus sparsiflorus* |
| Apache pine | *Pinus mayriana* |
| | *Pinus engelmannii* |
| apes earring | *Pithecellobium sonorae* |
| Arizona black walnut | *Juglans rupestris* |
| | *Juglans major* |
| Arizona cypress | *Cupressus arizonica* |
| Arizona grape | *Vitis arizonica* |

| | |
|---|---|
| Arizona madrone | *Arbutus arizonica* |
| Arizona morning glory | *Evolvulus arizonicus* |
| Arizona pine | *Pinus arizonica* |
| | *Pinus ponderosa* var *arizonica* |
| Arizona streptanthus | *Streptanthus arizonicus* |
| Arizona whiteoak | *Quercus arizonica* |
| arrowweed | *Pluchea sericea* |
| ashy jatropha | *Jatropha cinerea* |
| aspen fleabane | *Erigeron macranthus* |
| aster | *Aster* |
| | *Machaeranthera tanacetifolia* |
| Baja bursage | *Franseria peninsularis* |
| Baja candelilla | *Pedilanthus macrocarpus* |
| Baja cholla | *Opuntia cholla* |
| Baja lycium | *Lycium brevipes* |
| Baja ocotillo | *Fouquieria peninsularis* |
| banana yucca | *Yucca baccata* |
| barrel cactus | *Echinocactus polycephalus* |
| barrel cactus (biznaga) | *Echinocactus wislizeni* |
| | *Ferocactus wislizeni* |
| barreta | *Lindleyella mespiloides* |
| beavertail prickly pear | *Opuntia basilaris* |
| beefwood | *Casuarina* |
| bear grass (sacahuista) | *Nolina microcarpa* |
| big galleta | *Hilaria rigida* |
| Bigelow bear grass | *Nolina bigelovii* |
| Bigelow crossosoma | *Crossosoma bigelovii* |
| bitter cherry | *Cerasus crenulata* |
| | *Prunus emarginata* |
| black walnut | *Juglans rupestris* |
| bladder pod | *Isomeris arborea* |
| blue grama grass | *Bouteloua gracilis* |
| | *Bouteloua oligostachya* |
| blue palm | *Erythea armata* |
| blue paloverde | *Cercidium torreyanum* |
| | *Cercidium floridum* |
| boatthorn acacia | *Acacia cymbispina* |
| boojum tree (cirio) | *Idria columnaris* |
| borage | *Harpagonella* |
| boterboom | *Cotyledon paniculata* |
| brandega | *Brandegea bigelovii* |
| Brazil bloodwood tree | *Haemotoxylon brassiletto* |
| brittlebush | *Encilia farinosa* |
| broadbean | *Vicia faba* |

| | |
|---|---|
| broom weed | *Baccharis sarothroides* |
| buckhorn cholla | *Opuntia acanthocarpa* |
| buckwheat (wild) | *Eriognum abertianum* |
| buffalo gourd | *Cucurbita foetidissima* |
| bulb panic grass | *Panicum bulbosum* |
| burro brush | *Hymenoclea monogyra* |
| burroweed | *Haplopappus tenuisectus* |
| bursage | *Franseria* |
| bush mint | *Lippia* |
| | *Poliomintha incana* |
| bush muhly | *Muhlenbergia porteri* |
| bush redpepper | *Capsicum annuum* |
| California barrel cactus | *Ferocactus acanthodes* |
| California brickellia | *Brickellia californica* |
| | *Coleosanthus californicus* |
| California caltrop | *Kallstroemia californica* |
| California carlowrightia | *Carlowrightia californica* |
| California fagonia | *Fagonia californica* |
| California live oak | *Quercus agrifolia* |
| California lycium | *Lycium californicum* |
| California spectacle pod | *Dithyraea californica* |
| camelthorn | *Alhagi camelorum* |
| canaigre (wild dock) | *Rumex hymenosepalus* |
| candelilla | *Euphorbia antisyphilitica* |
| cane cholla | *Opuntia spinosior* |
| canutillo | *Ephedra californica* |
| canyon elderberry | *Sambucus vestita* |
| canyon live oak | *Quercus chrysolepis* |
| cardón | *Pachycereus calvus* |
| | *Pachycereus pringlei* |
| careless weed | *Amaranthus* |
| | *Amaranthus palmeri* |
| catclaw | *Acacia greggii* |
| catclaw mimosa | |
| (wait-a-minute bush) | *Mimosa biuncifera* |
| cattail | *Typha domingensis* |
| ceanothus (buck brush) | *Ceanothus* |
| chaenactis | *Chaenactis* |
| chamisa, (fourwing saltbush) | *Atriplex canescens* |
| chamise | *Adenostoma fasciculatum* |
| Chihuahua whitethorn | *Acacia vernicosa* |
| chihuahuan pine | *Pinus chihuahuana* |
| chirinola (creeping devil) | *Machaerocereus eruca* |

| | |
|---|---|
| chuparosa | *Beloperone californica* |
| cliff dogbane | *Apocynum scopulorum* |
| cliff fendlerbush | *Fendlera rupicola* |
| club cholla | *Opuntia clavata* |
| cocklebur | *Xanthium commune* |
| cochal | *Myrtillocactus trigonophylla* |
| coldenia | *Coldenia* |
| comandra | *Comandra pallida* |
| combseed | *Pectocarya* spp. |
| common nasturtium | *Tropaeoleum majus* |
| common reed | *Phragmites communis* |
| common sunflower | *Helianthus annuus* |
| copal | *Bursera hindsiana* |
| copper zephry lily | *Zephranthes longiflora* |
| coral bean | *Erythrina flabelliformis* |
| corkbark fir | *Abies arizonica* |
| | *Abies lasiocarpa* |
| Coulter lyre-fruit (mustard) | *Lyrocarpa coulteri* |
| coursetia | *Coursetia glandulosa* |
| Coville acuan | *Acuan covillii* |
| creeping devil (chirinola) | *Machaerocereus eruca* |
| creosote bush (gobernadora) | *Larrea tridentata* |
| | *Covillea tridentata* |
| | *Covillea* |
| crucifixion thorn | *Holacantha emoryii* |
| | *Canotia holacantha* |
| cryptantha | *Cryptanthe intermedia* |
| cudjoe wood | *Jacquinia pungens* |
| curved leaf agave | *Agave falcata* |
| cutleaf groundcherry | *Physalis angulata* |
| cypress pine | *Callitis* |
| dalea | *Parosela* |
| diamiana | *Turnera diffusa* |
| datilillo | *Yucca valida* |
| deervetch | *Lotus puberulus* |
| desert agave | *Agave deserti* |
| desert buckwheat | *Eriognum deserticola* |
| desert Christmas cactus | *Opuntia leptocaulis* |
| desert cotton | *Ingenhousia triloba* |
| | *Gossypium thurberi* |
| desert fir | *Peucephyllum schottii* |
| desert globemallow | *Sphaerelcea ambigua* |
| desert hackberry | *Celtis pallida* |

| | |
|---|---|
| desert holly | *Perezia wrightii* |
| | *Perezia nana* |
| desert hyacinth | *Brodiaea capitata* |
| | var *pauciflora* |
| | *Dichelostemma pulchellum* |
| | var *pauciflora* |
| desert Indian wheat | *Plantago fastigiata* |
| | *Plantago insularis* |
| desert lavender | *Hyptis emoryi* |
| desert lily (ajo) | *Hesperocallis undulata* |
| desert marigold | *Baileya multiradiata* |
| desert milkweed | *Asclepias subulata* |
| desert mistletoe | *Phoradendron californicum* |
| desert olive | *Forestiera neomexicana* |
| desert saltbush | *Atriplex polycarpa* |
| desert senna | *Cassia covesii* |
| desert thorn | *Lycium berlandieri* |
| desert tobacco | *Nicotiana trigonophylla* |
| desert willow | *Chilopsis linearis* |
| devils claw | *Martynia louisiana* |
| | *Proboscidea louisiana* |
| diamond cholla | *Opuntia ramosissima* |
| dogweed | *Dyssodia porophylloides* |
| Douglas fir | *Pseudotsuga mucronata* |
| | *Pseudotsuga taxifolia* |
| Douglas nightshade | *Solanum douglasii* |
| downy gilia | *Gilia floccosa* |
| drinn | *Aristida pungens* |
| dudleya | *Dudleya pulverulenta* |
| | *Dudleya arizonica* |
| Durango prickly pear | *Opuntia durangensis* |
| eardrop tree | *Enterolobium cyclocarpum* |
| elephant tree | *Bursera hindsiana* |
| | *Elaphrium macdougali* |
| | *Elaphrium microphyllum* |
| | *Bursera microphylla* |
| elkweed | *Frasera* |
| Emory bushmint | *Hyptis emoryi* |
| Emory oak | *Quercus emoryi* |
| Engelmann pricklypear | *Opuntia engelmannii* |
| | *Opuntia discata* |
| | *Opuntia phaeacantha* |
| Engelmann spruce | *Pieca engelmanii* |
| evening primrose | *Oenothera* |

| | |
|---|---|
| euphorbia (spurge) | *Euphorbia crenulata* |
| | *Euphorbia misera* |
| fairy duster | *Calliandra eriophylla* |
| false elephant tree | *Jatropha spathulata* |
| false tarragon sagebrush | *Artemisia dracunculoides* |
| fan palm | *Washingtonia filifera* |
| | *Washingtonia robusta* |
| feather pappusgrass | *Pappophorum wrightii* |
| Fendler ceanothus | *Ceanothus fendleri* |
| Fendler hedgehog cactus | *Echinocereus fendleri* |
| fetid marigold | *Pectis papposa* |
| fiddleneck | *Amsinckia tesselata* |
| finger-leaved gourd | *Cucurbita palmata* |
| fishhook cactus | *Mammillaria arizonica* |
| | *Mammillaria grahamia* |
| | *Mammillaria microcarpa* |
| flat-thorn acacia | *A. glabrata* |
| flattop buckwheat | *Eriognum fasciculatum* |
| fourwing saltbush | *Atriplex canescens* |
| foothill Indian wheat | *Plantago ignota* |
| foothill paloverde | *Cercidium microphyllum* |
| | *Parkinsonia microphylla* |
| four o'clock | *Mirabilis* |
| franseria (bursage) | *Acanthambrosia* |
| frankenia | *Frankenia* |
| Fremont cottonwood | *Populus fremontii* |
| Fremont pincushion | *Chaenactis* |
| Fremont lycium | *Lycium fremontii* |
| fried egg plant | *Argemone* |
| galleta | *Hilaria jamesii* |
| giant ragweed | *Ambrosia trifida* |
| gilia | *Gilia multiflora* |
| gland cinquefoil | *Potentilla glandulosa* |
| goldeneye | *Viguiera deltoidea* |
| goldenspined cactus, the lost cactus | *Pachycereus orcutii* |
| goldpoppy | *Eschscholtzia* |
| gordon bladderpod | *Lesquerella gordoni* |
| grama grass | *Bouteloua gracilis* |
| gray thorn | *Zizyphus lycioides* |
| | var *canescens* |
| | *Condalia lycioides* |
| | var *canescens* |
| green jointfir | *Ephedra virides* |
| Greenman partridgepea | *Chamaecrista leptadenia* |

| | |
|---|---|
| grevillea | *Grevillea* |
| grouse whortleberry | *Vaccinium scoparium* |
| guayule | *Parthenium argentatum* |
| hairy grama | *Bouteloua hirsuta* |
| hairbrush (cardón) | *Pachycereus pectenarboriginum* |
| hakea | *Hakea* |
| haloxylon | *Haloxylon* |
| hat-thorn acacia | *Acacia hindsii* |
| hawkweed | *Hieracium discolor* |
| | *Hieracium lemmoni* |
| hawthorn | *Crataegus* |
| heartleaf bursage | *Franseria cordifolia* |
| hedgehog cactus | *Echinocereus polyacanthos* |
| heterocentron | *Heterocentron* |
| hoary rosemary mint | *Poliomintha incana* |
| honey mesquite | *Prospopis juliflorañ* |
| | var *torreyana* |
| Hooker evening primrose | *Oenothera hookeri* |
| hop bush | *Dodonea viscosa* |
| | var *angustifolia* |
| hopsage | *Grayia spinosa* |
| horse purslane | *Tiranthema* |
| horse bean | *Parkinsonia aculeata* |
| horse mint | *Monarda pectinata* |
| horsebrush | *Tetradymia spinosa* |
| Humbolt coyotillo | *Karwinskia humboldtiana* |
| ibervillea | *Maximowiczia sonorae* |
| Indian mallow | *Abutilon* |
| interior live oak | *Quercus wislizeni* |
| iodine bush | *Spirostachys occidentalis* |
| | *Allenrolfea occidentalis* |
| ironwood (palo fiero) | *Olneya tesota* |
| Jerusalem thorn | *Parkinsonia aculeata* |
| Johnson echinomastus | *Echinomastus johnsonii* |
| joint fir (Mormon tea) | *Ephedra* spp |
| jojoba | *Simmondsia chinensis* |
| | *Simmondsia californica* |
| joshua tree | *Yucca brevifolia* |
| | *Yucca arborescens* |
| jujube | *Zizyphus* |
| jumping bean | *Sapium biloculare* |
| jumping cholla | *Opuntia fulgida* |
| kidney bean | *Phaseolus vulgaris* |
| kidney wood | *Eysenhardtia orthocarpa* |
| kopak tree | *Ceiba pentandra* |

krameria (ratany)

*Krameria glandulosa*
*Krameria parvifolia*
    var *glandulosa*

larchleaf goldenweed
    (turpentine weed)
laurel sumac
lechuguilla
lecidea
lignum vitae
limberbush
Lindheimer lipfern
little bluestem
little prickly pear
lotebush

*Chrysoma larcifolia*
*Haplopappus larcifolium*
*Rhus laurina*
*Agave lechuguilla*
*Lecidea*
*Guaiacum coulteri*
*Jatropha cardiophylla*
*Cheilanthes lindheimeri*
*Andropogon scoparius*
*Opuntia fragilis*
*Condalia lycioides*
*Condaliopsis lycioides*
    var *carescens*

Louisiana wormweed
madrone
mahonia barberry
maize-Indian corn
malefern
mallow
mallow, desert globemallow
mangle dulce (Baja)
manzanita
mariola
matacora
melaleuca
mellichampia
mescalito
mesquite
Mexican blue oak
Mexican crucillo

*Artemisia ludoviciana*
*Arbutus arizonica*
*Berberis wilcoxii*
*Zea mays*
*Aspidium filix-mas*
*Malva*
*Sphaeralcea ambigua*
*Maytenus phyllanthoides*
*Arctostaphylos pungens*
*Parthenium incanum*
*Jatropha cuneata*
*Melaleuca*
*Mellichampia ligulata*
*Hechtia montana*
*Prosopis*
*Quercus oblongifolia*
*Condalia spathulata*
*Condalia warnockii* var *kearney*

Mexican elderberry
Mexican grape
Mexican paloverde (horsebean)
Mexican pinyon pine
Mexican tea
Mexican whitebract
Mexican white pine
miguelito
milkvetch (locoweed)
Mojave yucca

*Sambucus mexicana*
*Cissus lacinata*
*Parkinsonia aculeata*
*Pinus cembroides*
*Ephedra trifurca*
*Hymenopappus mexicanus*
*Pinus strobiformis*
*Jatropha cordata*
*Astragalus*
*Yucca schidigera*

| | |
|---|---|
| mountainrose | *Antigonon leptopus* |
| mountain snowberry | *Symphoricarpos oreophilus* |
| mourita | *Lysiloma divaricata* |
| mortonia (sandpaper bush) | *Mortonia scabrella* |
| muhly (purple muhly) | *Muhlenbergia affinis* |
| | *Muhlenbergia rigida* |
| mulberry | *Morus celtidifolia* |
| myoporum | *Myoporum* |
| narrowleaf cottonwood | *Populus angustifolia* |
| netleaf hackberry | *Celtis reticulata* |
| netleaf oak | *Quercus reticulata* |
| New Mexico flax | *Linum neomexicanum* |
| New Mexico groundsel | *Senecio neomexicanus* |
| New Mexico locust | *Robina neomexicana* |
| Nevada squaw tea | *Ephedra nevadensis* |
| nievitas | *Eritrichium* sp. |
| | *Cryptantha* sp. |
| nightblooming cereus | *Cereus greggii* |
| ocotillo | *Fouquieria splendens* |
| one-seed juniper | *Juniperus monosperma* |
| orange gooseberry | *Ribes pinetorum* |
| organpipe cactus | *Cereus thurberi* |
| | *Lemairiocereus thurberi* |
| owlclover | *Orthocarpus purpurascens* |
| palafoxia | *Palafoxia* |
| Palmer agave | *Agave palmeri* |
| Palmer penstemon | *Penstemon palmeri* |
| palmilla | *Yucca elata* |
| | *Yucca radiosa* |
| palo blanco | *Lysiloma candida* |
| palo brea | *Cercidium praecox* |
| palo San Juan | *Forchameria watsonii* |
| palo zorillo | *Cassia emarginata* |
| pancake prickly pear | *Opuntia chlorotica* |
| paper bag bush | *Salazaria mexicana* |
| paper flower | *Psilostrophe cooperi* |
| Parry penstemon | *Penstemon parryi* |
| patata | *Monolepis nuttaliana* |
| pencil cholla | *Opuntia arbuscula* |
| | *Opuntia kleinae* |
| | *Opuntia thurberi* |
| pickleweed | *Allenrolfea* |
| pigweed | *Trianthema portulacastrum* |
| pinyon pine | *Pinus edulis* |
| pitahaya agria | *Machaerocereus gummosus* |

| plains prickly pear | Opuntia polycantha |
|---|---|
| pochote | Ceiba acuminata |
| ponderosa pine | Pinus ponderosa |
| prairie acacia | Acacia suffrutenscens |
| | Acacia angustissima |
| pricklypear | Opuntia spp. |
| prickly poppy (fried egg plant) | Argemone spp. |
| pussytoes | Antennaria marginata |
| pyrola | Pyrola sp. |
| quaking aspen | Populus tremuloides |
| queen's wreath | Antigonon leptopus |
| quivertree aloe | Aloe dichotoma |
| range krameria | Krameria glandulosa |
| redberry buckthorn | Rhamnus ilicifolia |
| | Rhamnus crocea |
| red molly | Kochia americana |
| redosier dogwood | Cornus stolonifera var riparia |
| redstar zinnia | Crassina pumila |
| | Zinnia acerosa |
| ring muhly | Muhlenbergia gracillima |
| | Muhlengergia torreyi |
| rose mallow | Hibiscus denudatus |
| rosetree | Vauquelinia californica |
| Rothrox grama | Bouteloua rothrockii |
| rough menodora | Menodora scabra |
| rubus | Rubus arizonicus |
| sagebrush | Artemisia |
| sagebrush (big) | Artemisia tridentata |
| saguaro | Carnegiea gigantea |
| | Cereus giganteus |
| saltgrass | Distichlis spicata |
| samandoca | Yucca carnerosa |
| | Samuela carnerosa |
| San Diego redberry buckthorn | Rhamnus crocea var pilosa |
| sandpaper plant | Petallonyx thurberi |
| sand peppergrass | Lepidium lasiocarpum |
| sand verbena | Abronia villosa |
| sangre de drago (limberbush) | Jatropha cordata |
| Santa Rita prickly pear | Opuntia santa rita |
| Schotts yucca | Yucca macrocarpa |
| | Yucca schottii |
| screw bean | Prosopis pubescens |
| seepweed | Suaeda |
| scrub oak | Quercus turbinella |
| seepwillow baccharis | Baccharis glutinosa |

| | |
|---|---|
| selaginella | *Selaginella rupicola* |
| senita (sinita) | *Cereus schotii* |
| | *Lophocereus schottii* |
| shadscale | *Atriplex confertifolia* |
| showy goldeneye | *Gymnolomia multiflora* |
| shrub nightshade | *Solanum hindsianum* |
| shrubby buckwheat | *Eriogonum wrightii* |
| sideoats grama | *Bouteloua curtipendula* |
| silk tassel | *Garrya wrightii* |
| silver bluestem | *Andropogon saccharoides* |
| silver cholla | *Opuntia echinocarpa* |
| silver lotus | *Anisolotus argensis* |
| | *Lotus rigidus* |
| silver leaf oak | *Quercus hypoleucoides* |
| sixweeks fescue | *Festuca octoflora* |
| sixweeks needle grass | *Bouteloua aristidoides* |
| skunk bush sumac | *Rhus trilobata* |
| slender alliona | *Allionia gracillima* |
| slender poreleaf (hierba de venada) | *Porophyllum gracile* |
| slim combseed | *Pectocarya linearis* |
| slim pod senna | *Cassia leptocarpa* |
| slim tridens | *Triodia mutica* (obs) |
| | *Tridens multicus* |
| small leaf ayenia | *Ayenia microphylla* |
| smoke tree | *Dalea spinosa* |
| smooth colubrina | *Colubrina glabra* |
| smooth prickly pear | *Opuntia leavis* |
| sneezeweed | *Dugaldia hoopesii* |
| | *Helenium hoopesii* |
| soapberry | *Sapindus saponaria* |
| | var *drummondii* |
| Sonora jumping cholla | *Opuntia mammillata* |
| Sonoran caper | *Atamisquea emarginata* |
| Sonoran cordia | *Cordia sonorae* |
| Sonoran croton | *Croton sonorae* |
| Sonora palmetto | *Sabal uresana* |
| Sonora paloverde | *Cercidium sonorae* |
| Sonora rathbun cactus | *Rathbunia alamosensis* |
| southwestern chokecherry | *Prunus virens* |
| soyate | *Beaucarnea inermis* |
| spanish needle | *Palafoxia linearis* |
| spidergrass | *Aristida ternipes* |
| spiderling | *Boerhaavia intermedia* |
| spindletree | *Fusanus* sp. |
| spiny allthorn | *Koeberlinia spinosa* |

| | |
|---|---|
| spurge | *Euphorbia* |
| staghorn cholla | *Opuntia versicolor* |
| star cloakfern | *Notholaena hookeri* |
| steel acacia | *Acacia macracantha* |
| stonecrop | *Sedum stenopetalum* |
| sulfur prickly pear | *Opuntia streptacantha* |
| sweet acacia | *Acacia farnesiana* |
| sweetbush | *Bebbia juncea* |
| sycamore | *Platanus wrightii* |
| tamarisk | *Tamarix* spp. |
| tanglehead | *Heteropogon contortus* |
| tar bush | *Flourensia cernua* |
| tarweed | *Hemizonia* |
| tasselflower | *Brickellia grandiflora* |
| tea tree | *Leptospermum* |
| teasel | *Dipsacus* |
| teddy bear cholla | *Opuntia bigelovii* |
| teso | *Acacia occidentalis* |
| tesota (ironwood) | *Olneya tesota* |
| Texas algerita | *Mahonia trifoliolata* |
| Texas croton | *Croton texensis* |
| threeawn | *Aristida* spp. |
| three seeded lotus | *Anisolotus trispermus* |
| | *Lotus humistratus* |
| | *Anisolotus argensis* |
| Thurber anisacanthus | *Anisacanthus thurberi* |
| (desert honeysuckle) | |
| Thurber sandpaper plant | *Petalonyx thurberi* |
| tidestromia | *Tidestromia* spp. |
| tillandsia (heno pequeno, bullmoss) | *Tillandsia recurvata* |
| tobosa grass | *Hilaria mutica* |
| tomatillo | *Solanum elaeagnifolium* |
| torote | *Bursera microphylla* |
| | *Bursera filicifolia* |
| | *Bursera odorata* |
| | *Bursera latiflora* |
| Toumey prickly pear | *Opuntia toumeyi* |
| tree cholla | *Opuntia imbricata* |
| tree ocotillo | *Fouquieria macdougalii* |
| tree morning glory | *Ipomoea arborescens* |
| | *Ipomoea murucoides* |
| triangle leaf bursage | *Franseria deltoidea* |
| trixis | *Trixis angustifolia* |
| | var *angustifolia* |
| tuber anemone | *Anemone tuberoso* |

| | |
|---|---|
| tuna prickly pear | *Opuntia tuna* |
| twin flower | *Janusia gracilis* |
| twisted acacia | *Acacia tortuosa* |
| Utah juniper | *Juniperus osteosperma* |
| Utah white oak | *Quercus submollis* |
| | *Quercus utahensis* |
| velvet mesquite | *Prosopis juliflora* var *velutina* |
| velvet pod mimosa | *Mimosa dysacarpa* |
| verbena | *Verbena* spp. |
| violet phacelia | *Phacelia distans* |
| Virginia creeper | *Parthenocissus vitacea* |
| | *Parthenocissus inserta* |
| viscainoa | *Viscainoa geniculata* |
| welwitschia | *Welwitschia mirabilis* |
| western bracken | *Pteris aquilina* |
| western honey mesquite | *Prosopis juliflora* var *torreyana* |
| western virginsbower | *Clematis ligusticifolia* |
| wheat | *Triticum* |
| Wheeler sotol | *Dasylirion wheeleri* |
| white dalea | *Dalea emoryi* |
| white bursage | *Franseria dumosa* |
| white fir | *Abies concolor* |
| whitestem mentzelia | *Mentzelia albicaulis* |
| whitethorn | *Acacia constricta* |
| wholeleaf painted cup | *Castilleja integra* |
| wild buckwheat | *Eriogonum* |
| wild carrot | *Daucus pusillus* |
| wild heliotrope | *Phacelia crenulata* |
| wild hollyhock | *Sphaeralcea pedata* |
| | *Sphaeralcea grossularifolia* |
| Willard acacia | *Acacia willardiana* |
| willow | *Salix* spp. |
| willow groundsel | *Senecio salignus* |
| winter fat | *Eurotia lanata* |
| woolly butterflybush | *Buddleya marrubiifolia* |
| Whipple cholla | *Opuntia whipplei* |
| Wright bean | *Phaseolus wrightii* |
| Wright bush mint | *Lippia wrightii* |
| | *Aloysia wrightii* |
| Wright prickly pear | *Opuntia wrightiana* |
| Wrights penstemon | *Penstemon wrightii* |
| yellow felt plant | *Horsfordia newberryi* |
| yellow trumpet | *Tecoma stans, Stenolobium stans* |
| yerba de pasmo | *Baccharis pteronioides* |
| zygophyllum | *Zygophyllum* spp. |

## Scientific Names–Common Names

| | |
|---|---|
| *Abies arizonica* (syn)<br>   see *Abies lasiocarpa* | corkbark fir |
| *Abies concolor* | white fir |
| *Abies lasiocarpa* | corkbark fir |
| *Abronia vilosa* | sand verbena |
| *Abutilon* spp. | Indian mallow |
| *Acacia constricta* | whitethorn |
| *Acacia cymbispina* | boatthorn acacia |
| *Acacia farnesiana* | sweet acacia |
| *Acacia glabrata* | flatthorn acacia |
| *Acacia greggii* | catclaw |
| *Acacia hindsii* | hat-thorn acacia |
| *Acacia macracantha* | steel acacia |
| *Acacia occidentalis* | teso |
| *Acacia suffretescens* | prairie acacia |
| *Acacia tortuosa* | twisted acacia |
| *Acacia vernicosa* | Chichuhua whitethorn |
| *Acacia willardiana* | Willard acacia |
| *Acanthambrosia* (syn) | franseria (bursage) |
| *Acarospora* | acarospora |
| *Acuan covillei* (syn)<br>   see *Desmanthus covillei* | Coville's acuan |
| *Adenostoma fasciculatum* | chamise |
| *Agave deserti* | desert agave |
| *Agave falcata* | curved leaf agave |
| *Agave lechuguilla* | lechuguilla |
| *Agave palmeri* | Palmer agave |
| *Agave schottii* | amole |
| *Alhagi camleorum* | camelthorn |
| *Allenrolfea occidentalis* | pickleweed (iodine bush) |
| *Allionia gracillima* | slender alliona |
| *Alnus* spp. | alder |
| *Aloe dichotoma* | quivertree aloe |
| *Amaranthus palmeri* | careless weed |
| *Ambrosia*<br>   see *Franseria* | |
| *Ambrosia trifida* | giant ragweed |
| *Amsinckia tesselata* | fiddleneck |
| *Andropogon saccharoides* | silver bluestem |

| | |
|---|---|
| *Andropogon scoparius* | little bluestem |
| *Anemone tuberosa* | tuber anemone |
| *Anisolotus argensis* (syn) | silver lotus |
| see *Lotus rigidus* | |
| *Anisacanthus thurberi* | Thurber anisacanthus |
| *Anisolotus trispermus* (syn) | three-seeded lotus |
| see *Lotus humistratus* | |
| see *Anisolotus argensis* | |
| *Antennaria marginata* | pussytoes |
| *Antigonon leptopus* | queen's wreath (mountainrose) |
| *Apocynum scopulorum* | cliff dogbone |
| *Arbutus arizonica* | Arizona madrone |
| *Arctostaphylos pungens* | manzanita |
| *Argemone* spp. | prickly poppy |
| | (fried egg plant) |
| *Aristida* spp. | threeawn grass |
| *Aristida pungens* | drinn |
| *Artemisia* spp. | sagebrush |
| *Artemisia dracunculoides* | false tarragon sagebrush |
| *Artemisia ludoviciana* | Louisiana wormwood |
| *Artemisia tridentata* | sagebrush (big) |
| *Asclepias subulata* | desert milkweed |
| *Aspidium filix-mas* | malefern |
| *Aster* spp. | aster |
| *Astragalus* spp. | milkvetch (loco weed) |
| *Atamisquea emarginata* | Sonoran caper |
| *Atriplex canescens* | chamisa, fourwing saltbush |
| *Atriplex confertifolia* | shadscale |
| *Atriplex polycarpa* | desert salt bush |
| *Ayenia microphylla* | small leaf ayenia |
| *Baccharis pteronioides* | yerba de pasmo |
| *Baccharis sarothroides* | broom weed |
| *Baccharis thesioides* | no common name |
| *Baileya multiradiata* | desert marigold |
| *Beaucarnea inermis* | soyate |
| *Bebbia juncea* | sweetbush |
| *Beloperone californica* | chuparosa |
| *Berberis wilcoxii* | mahonia barberry |
| *Bigelovia hartwegii* (syn) | burroweed |
| see *Haplopappus tenuisectus* | |
| *Beorhaavia intermedia* | spiderling |
| *Bouteloua aristidoides* | sixweeks needle grass |
| *Bouteloua curtipendula* | sideoats grama |
| *Bouteloua gracilis* | blue grama grass |
| *Bouteloua hirsuta* | hairy grama |

| | |
|---|---|
| *Bouteloua oligostachya* (syn) | blue grama grass |
| see *Bouteloua gracilis* | |
| *Bouteloua rothrockii* | Rothrox grama |
| *Brandegea bigelovii* | brandega |
| *Brickellia californica* | California brickellia |
| see *Coleosanthus californicus* | |
| *Brickellia grandiflora* | tassleflower brickellia |
| *Brodiaea capitata* | desert hyacinth |
| *Brodiaea capitata* var *pauciflora* | desert hyacinth |
| see *Dichelostemma pulchellum* | |
| var *pauciflorum* | |
| *Bromus richardsonii* | brome grass |
| *Buddleya marrubiifolia* | woolly butterflybush |
| *Bursera hindiana* | copal |
| *Bursera latiflora* | torote |
| *Bursera microphylla* | torote |
| *Bursera odorata* | torote |
| *Callidandra eriophylla* | fairy duster |
| *Callitus* sp. | cypress pine |
| *Canotia holacantha* | crucifixion thorn |
| *Capsicum annuum* | bush redpepper |
| *Carlowrightia californica* | California carlowrightia |
| *Carnegiea gigantea* | saguaro |
| *Cassia covesii* | desert senna |
| *Cassia emarginata* | palo zorillo |
| *Cassia loptocarpa* | slim pod senna |
| *Castilleja integra* | wholeleaf painted cup |
| *Casuarina* | beefwood |
| *Ceanothus* spp. | ceanothus |
| *Ceanothus fendleri* | Fendler ceanothus |
| *Ceiba acuminata* | pochote |
| *Ceiba pentandra* | kopak tree |
| *Celtis pallida* | desert hackberry |
| *Celtis reticulata* | netleaf hackberry |
| *Cerasus crenulata* | bitter cherry |
| see *Prunus emarginata* | |
| *Cercicium microphyllum* | foothill paloverde |
| *Cercidium praecox* | palo brea |
| *Cercidium sonorae* | sonora paloverde |
| *Cercidium terreyanum* | blue paloverde |
| *Cereus giganteus* (syn) | sahuaro |
| see *Carnegiea gigantea* | |
| *Cereus greggii* (syn) | nightblooming cereus |
| see *Peniocereus greggii* | |

*Cereus schottii*                                   senita (sinita) (garumbillo)
*Cereus thurberi*
  see *Lemairiocereus thurberi*                     organpipe cactus
*Chaenactis*                                        Freemont pincushion
*Chamaecrista leptadenia*                           Greenman partridgepea
*Cheilanthes lindheimeri*                           Lindheimer lipfern
*Chilopsis linearis*                                desert willow
*Chrysoma laricifolius*                             larchleaf goldenweed
  see *Haplopappus laricifolius*
*Cissus lacinata*                                   Mexican grape
*Clematis ligusticifolia*                           western virginsbower
*Coldenia* sp.                                      coldenia
*Coleosanthus californicus* (syn)                   California brickellia
  see *Brickellia californica*
*Colubrina glabra*                                  smooth colubrina
*Comandra pallida*                                  comandra
*Condalia lycioides*                                lotebush
*Condalia lycioides* var *canescens*                gray thorn
*Condalia spathulata*                               Mexican crucillo
*Cordia boissieri*                                  anacahuita
*Cordia sonorae*                                    Sonoran cordia
*Cordylanthus wrightii*                             Wright birdbeak
*Cornus stolonifera* var *riparia*                  redosier dogwood
*Cotyledon panicalata*                              boterboom
*Coursetia glandulosa*                              coursetia
*Covillea*                                          creosote bush
  see *Larrea*
*Covillea tridentata* (syn)                         creosote bush (gobernadora)
  see *Larrea tridentata*
*Crassina pumila* (syn)                             redstar zinnia
  see *Zinnia acerosa*
*Crataegus*                                         hawthorn
*Crossosoma bigelovii*                              Bigelow crossosoma
*Croton sonorae*                                    Sonora croton
*Croton texensis*                                   Texas croton
*Cryptantha intermedia*                             cryptantha
*Cucurbita foetidissima*                            buffalo gourd
*Cucurbita palmata*                                 finger-leaved gourd
*Cupressus arizonica*                               Arizona cypress
*Cynandum sinaloensis*                              mellichampia
*Dalea emoryi*                                      white dalea
*Dalea spinosa*                                     smoke tree
*Dasylirion wheeleri*                               Wheeler sotol
*Daucus pusillus*                                   wild carrot

| | |
|---|---|
| *Dichelostemma pulchellum* var *pauciflorum* (syn) see *Brodisea capitata* | desert hyacinth |
| *Dipsacus* sp. | teasel |
| *Distichlis spicata* | saltgrass |
| *Dithyraea californica* | California spectacle pod |
| *Dodonaea viscosa* var *augustifolia* | hop bush |
| *Dugaldia hoopesii* (syn) see *Helenium hoopesii* | sneezeweed |
| *Dudleya pulverulenta* | Dudleya |
| *Dyssodia porophylloides* | dogweed |
| *Echinocactus polycephalus* | barrel cactus |
| *Echinocereus fendleri* | fendler hedgehog cactus |
| *Echinocereus polyacanthos* | hedgehog cactus |
| *Echinocactus wislizeni* (syn) see *Ferocactus wislizeni* | barrel cactus (ferocactus) |
| *Echinomastus johnsonii* | Johnson echinomastus |
| *Elaphrium microphyllum* (syn) see *Bursera microphylla* | elephant tree |
| *Elaphrium macdougali* (syn) see *Bursera hindsiana* | elephant tree or copal |
| *Encelia farinosa* | brittlebush |
| *Enterolobium cyclocarpum* | eardrop tree |
| *Ephedra* spp. | joint fir |
| *Ephedra californica* | canutillo |
| *Ephedra nevadensis* | Nevada squaw tea |
| *Ephedra trifurca* | Mexican tea |
| *Ephedra virides* | green joint fir |
| *Eremophila* | |
| *Erigeron macranthus* | aspen fleabane |
| *Eritrichium* sp. see *Cryptantha* sp. | nievitas |
| *Eriocarpus gracile* | annual haplopappus (aplopappus) |
| *Eriognum abertianum* | buckwheat (wild) |
| *Eriognum diserticola* | desert buckwheat |
| *Eriognum fasciculatum* | flattop buckwheat |
| *Eriognum wrightii* | shrubby buckwheat |
| *Erodium* | filaree |
| *Erodium cicutarium* | filaree (afileria) |
| *Erythea armata* | blue palm |
| *Erythrina flabelliformis* | coral bean |
| *Eschscholtzia* spp. | goldpoppy |
| *Eucalyptus* spp. | eucalyptus |

| | |
|---|---|
| *Euphorbia* spp. | spurge |
| *Euphorbia antisyphilitica* | candelilla |
| *Euphorbia crenulata* | euphorbi, or spurge |
| *Euphorbia misera* | spurge |
| *Eurotia lanata* | winter fat |
| *Evolvulus arizonicus* | Arizona morning glory |
| *Eysenhardtia orthocarpa* | kidney wood |
| *Fagonia californica* | California fagonia |
| *Fendlera rupicola* | cliff fendlerbush |
| *Ferocactus acanthodes* | California barrel cactus |
| *Ferocactus wislizeni* | barrel cactus |
| *Festuca octoflora* | six weeks fescue |
| *Flourensia cernua* | tar bush |
| *Forchammeria watsonii* | palo San Juan |
| *Forestiera neomexicana* | desert olive |
| *Fouquieria macdougalii* | tree ocotillo |
| *Fouquieria peninsularis* | Baja ocotillo |
| *Fouquieria splendens* | ocotillo |
| *Frankenia* | frankenia |
| *Franseria* | bursage |
| see *Ambrosia* | |
| *Franseria ambrosioides* | ambrosia bursage |
| *Franseria chenopodifolia* | San Diego bursage |
| *Franseria cordifolia* | heartleaf bursage |
| *Franseria deltoidea* | triangle leaf bursage |
| *Franseria dumosa* | white bursage |
| *Franseria peninsularis* | Baja bursage |
| *Frasera* | elkweed |
| *Fusanus* | spindletree |
| *Garrya wrightii* | silk tassel |
| *Gilia floccosa* | downy gilia |
| *Gilia multiflora* | gilia |
| *Gossypium thurberi* | desert cotton |
| *Grayia spinosa* | hopsage |
| *Grevillea* | grevillea |
| *Guaiacum coulteri* | lignum vitae |
| *Gymnolomia multiflora* | showy goldeneye |
| *Haemotoxylon brasiletto* | Brazil bloodwood tree |
| *Hakea* | hakea |
| *Haloxylon* | haloxylon |
| *Haplopappus hartwegii* | burroweed |
| *Haplopappus laricifolius* | larchleaf goldenweed (turpentine weed) |
| *Haplopappus tenuisectus* | burroweed |
| *Harpagonella* | borage |

| | |
|---|---|
| *Hechtia montana* | mescalito |
| *Helenium hoopesii* | sneezeweed |
| *Helianthus annuus* | common sunflower |
| *Hemizonia* | tarweed |
| *Hesperocallis undulata* | desert lily (ajo lily) |
| *Heteropogon contortus* | tanglehead |
| *Hibiscus denudatus* | rose mallow |
| *Hieracium discolor* | hawkweed |
| *Hieracium lemmoni* | hawkweed |
| *Hilaria jamesi* | galleta |
| *Hilaria mutica* | tobosa grass |
| *Hilaria rigida* | big galleta |
| *Holacantha emoryi* | crucifixion thorn |
| *Hordeum murinum* | cheat grass |
| *Horsfordia newberryi* | yellow felt plant |
| *Hymenopappus mexicanus* | Mexican whitebract |
| *Hymenoclea monogyra* | burro brush |
| *Hyptis emoryi* | Emory bushmint |
| *Idria columnaris* | boojum tree (cirio) |
| *Ingenhousia triloba* (syn) | desert cotton |
| see *Gossypium thruberi* | |
| *Ipomoea arborescens* | tree morning glory |
| *Isocoma hartwegi* (syn) | burroweed |
| see *Haplopappus hartwegii* | |
| *Isomeris arborea* | bladder pod |
| *Jacquinia pungens* | cudjoe wood |
| *Janusia gracilis* | twinflower |
| *Jatropha cardiophylla* | sangre de drago (limberbush) |
| *Jatropha cinerea* | ashy jatropha |
| *Jatropha cordata* | miguelito |
| *Jatropha cuneata* | matacora |
| *Jatropha spathulata* | false elephant tree |
| *Juniperus deppeana* | alligator juniper |
| *Juniperus monosperma* | one-seed juniper |
| *Juniperus osteosperma* | Utah juniper |
| *Juniperus pachyphloea* (syn) | alligator juniper |
| see *Juniperus deppeana* | |
| *Juglans rupestris* | black walnut |
| *Kallstroemia californica* | California caltrop |
| *Karwinskia humboldtiana* | Humbolt coyotillo |
| *Kochia americana* | red molly |
| *Koeberlinia spinosa* | spiny allthorn |
| *Krameria glandulosa* | krameria (range ratany) |
| *Larrea tridentata* | creosote bush (gobernadora) |
| *Lemairiocereus thurberi* | organ-pipe cactus |

| | |
|---|---|
| *Lepidium lasiocarpum* | sand peppergrass |
| *Leptospermum* | tea tree |
| *Lesquerella gordoni* | Gordon bladderpod |
| *Lindleyella mespiloides* | baretta |
| *Linum neomexicanum* | New Mexico flax |
| *Lippa* | bushmint |
| *Lippia wrightii* | Wright bushmint |
| *Lophocereus schottii* | senita (garambullo) |
| *Lotus humistratus* (syn) | silver lotus |
| see *Anisolatus argensis* | |
| *Lotus puberulus* | deervetch |
| *Lupinus sparsiflorus* | annual lupine |
| *Lycium andersonii* | Anderson lycium |
| *Lycium berlandieri* | desert thorn |
| *Lycium brevipes* | Baja lycium |
| *Lycium californicum* | California lycium |
| *Lycium fremontii* | Fremont lycium |
| *Lyrocarpa coulteri* | Coulter lyre-fruit (mustard) |
| *Lysiloma candida* | palo blanco |
| *Lysiloma divaricata* | mourita |
| *Machaeranthera tanacetifolia* | aster |
| *Machaerocereus eruca* | creeping devil (chirinola) |
| *Machaerocereus gummosus* | pitahaya agria |
| *Mahonia trifoliolata* | Texas algerita |
| *Malva* | mallow |
| *Mammillaria arizonica* | fishhook cactus |
| *Mammillaria grahami* | fishhook cactus |
| *Mammillaria microcarpa* | fishhook cactus |
| *Martynia louisiana* | devils claw |
| *Maximowiczia sonorae* | ibervillea |
| *Maytenus phyllanthoides* | mangle dulce |
| *Melaleuca* | melaleuca |
| *Mellichampia ligulata* (syn) | mellichampia |
| see *Cynandum sinaloensis* | |
| *Menodora scabra* | rough mendora |
| *Mentzelia albicaulis* | whitestem mentzelia |
| *Mimosa biuncifera* | catclaw mimosa (wait-a-minute bush) |
| *Mimosa dysacarpa* | velvet pod mimosa |
| *Mirabilis* | four o'clock |
| *Monarda pectinata* | horsemint |
| *Monolepis nuttaliana* | patata |
| *Mortonia scabrella* | sandpaper bush |
| *Morus microphylla* | mulberry |

| *Muhlenbergia affinis* (syn) | muhly |
| see *Muhlenbergia rigida* | |
| *Muhlenbergia gracillima* (syn) | ring muhly |
| see *Muhlenbergia torreyi* | |
| *Muhlenbergia porteri* | bush muhly |
| *Muhlenbergia rigida* | muhly |
| *Muhlenbergia torreyi* | ring muhly |
| *Myoporum* | myoporum |
| *Myrtillocactus cochal* | cochal |
| *Nicotiana trigonophylla* | desert tobacco |
| *Nolina bigelovii* | Bigelow bear grass |
| *Nolina microcarpa* | beargrass (sacahuista) |
| *Notholaena hookeri* | star cloakfern |
| *Oenothera* | evening primrose |
| *Oenothera hookeri* | Hooker evening primrose |
| *Olneya tesota* | ironwood (palo fiero) |
| *Opuntia arbuscula* | pencil cholla |
| *Opuntia basilaris* | beavertail prickly pear |
| *Opuntia bigelovii* | teddy bear cholla |
| *Opuntia blakeana* | prickly pear |
| *Opuntia chlorotica* | pancake prickly pear |
| *Opuntia cholla* | Baja cholla |
| *Opuntia clavata* | club cholla |
| *Opuntia discata* | Engelmann prickly pear |
| *Opuntia durangensis* | Durango prickly pear |
| *Opuntia echinocarpa* | silver cholla |
| *Opuntia engelmannii* | Engelmann prickly pear |
| *Opuntia fragilis* | little prickly pear |
| *Opuntia fulgida* | jumping cholla |
| *Opuntia imbricata* | tree cholla |
| *Opuntia kleinae* | pencil cholla |
| *Opuntia laevis* | smooth prickly pear |
| *Opuntia leptocaulis* | desert Christmas cactus |
| *Opuntia mammillata* | (Sonora) jumping cholla |
| *Opuntia phaeacantha* | Engelmann prickly pear |
| *Opuntia ramosissima* | diamond cholla |
| *Opuntia santa rita* | Santa Rita prickly pear |
| *Opuntia spinosior* | cane cholla |
| *Opuntia streptacantha* | sulfur prickly pear |
| *Opuntia thurberi* | pencil cholla |
| *Opuntia toumeyi* | Toumey prickly pear |
| *Opuntia tuna* | tuna prickly pear |
| *Opuntia versicolor* | staghorn cholla |
| *Opuntia wrightiana* | Wright Prickly pear |

| | |
|---|---|
| *Opuntia whipplei* | Whipple cholla |
| *Orthocarpus prupurascens* | owlclover |
| *Pachycereus calvus* (syn) | cardón |
| see *Pachycereus pringlei* | |
| *Pachycereus orcutii* | golden spined cactus, the lost cactus |
| *Pachycereus pecten-aboriginum* | hairbrush (cactus) cardón |
| *Pachycereus pringlei* | cardón |
| *Pachycormus discolor* | elephant tree |
| *Palafoxia* | spanish needle |
| *Palma samandoca* (syn) | |
| see *Samuela carnerosa* | samandoca |
| *Panicum bulbosum* | bulb panic grass |
| *Pappophorum wrightii* | feather pappusgrass |
| *Parkinsonia* (syn) | Mexican paloverde |
| *Parkinsonia aculeata* | Mexican paloverde - horsebean - Jerusalem thorn |
| *Parkinsonia microphylla* (syn) | foothill paloverde |
| see *Cercidium microphyllum* | |
| *Parosela* | dalea |
| *Parthenium argentatum* | guayule |
| *Parthenium incanum* | mariola |
| *Parthenocissus vitacea* | Virginia creeper |
| *Pectus papposa* | fetid marigold |
| *Pectocarya* | combseed |
| *Pectocarya linearis* | slim combseed |
| *Pedilanthus macrocarpus* | Baja candelilla |
| *Peniocereus greggii* | nightblooming cereus |
| *Penstemon palmeri* | Palmer penstemon |
| *Penstemon parryi* | Parry penstemon |
| *Penstemon wrightii* | Wright penstemon |
| *Perezia nana* | desert holly |
| *Perezia wrightii* | desert holly |
| *Petalonyx thurberi* | sandpaper plant |
| *Peucephyllum schottii* | desert fir |
| *Phacelia crenulata* | wild heliotrope |
| *Phacelia distans* | violet phacelia |
| *Phaseolus vulgaris* | kidney bean |
| *Phaseolus wrightii* | Wright bean |
| *Phoradendron californicum* | desert mistletoe |
| *Phragmites communis* | common reed |
| *Physalis angulata* | cutleaf groundcherry |
| *Picea engelmannii* | Engelmann spruce |
| *Pinus arizonica* (syn) | Arizona pine |
| see *Pinus ponderosa* var *arizonica* | |

| | |
|---|---|
| *Pinus cembroides* | Mexican pinyon pine |
| *Pinus chihuahuana* | chihuahuan pine |
| *Pinus edulis* | pinyon pine |
| *Pinus engelmannii* | Apache pine |
| *Pinus mayriana* (syn) | Apache pine |
|   see *Pinus engelmannii* | |
| *Pinus ponderosa* | ponderosa pine |
| *Pinus strobiformis* | Mexican white pine |
| *Pithecellobium sonorae* | apes earring |
| *Plantago ignota* | foothill Indian wheat |
| *Plantago fastigiata* | desert Indian wheat |
| *Plantago insularis* | desert Indian wheat |
| *Plantanus wrightii* | sycamore |
| *Pluchea sericea* | arrowweed |
| *Poliomintha incana* | bush mint |
| *Populus angustifolia* | narrowleaf cottonwood |
| *Populus fremontii* | cottonwood |
| *Populus tremuloides* | quaking aspen |
| *Porophyllum gracile* | slender poreleaf |
| |   (hierba de venado) |
| *Potentilla glandulosa* | gland cinquefoil |
| *Prosopis* | mesquite |
| *Prosopis juliflora* | |
|   var *torreyana* | western honey mesquite |
| *Prosopis pubescens* | screw bean |
| *Prosopis juliflora* | velvet mesquite |
|   var *velutina* | |
| *Prunus emarginata* | bittercherry |
| *Prunus virens* | southwestern chokecherry |
| *Psilostrophe cooperi* | paper flower |
| *Pseudotsuga mucronota* (syn) | douglas fir |
| *Pseudotsuga taxifolia* | douglas fir |
| *Pteris aquilina* var *pubescens* | western bracken |
| *Pyrola* | pyrola |
| *Quercus agrifolia* | California live oak |
| *Quercus arizonica* | Arizona white oak |
| *Quercus chrysolepis* | canyon live oak |
| *Quercus emoryi* | Emory oak |
| *Quercus hypoleucoides* | silverleaf oak |
| *Quercus oblongifolia* | Mexican blue oak |
| *Quercus reticulata* | netleaf oak |
| *Quercus submollis* | Utah white oak |
| *Quercus turbinella* | scrub oak |
| *Quercus wislizeni* | interior live oak |
| *Randia thurberi* | Thurber randia |

| | |
|---|---|
| *Rathbunia alamosensis* | Sonora rathbun cactus |
| *Rhamnus crocea* var *pilosa* | San Diego redberry buckthorn |
| *Rhamnus ilicifolia* | redberry buckthorn |
| *Rhizophora mangle* | American mangrove |
| *Rhus laurina* | laurel sumac |
| *Rhus trilobata* | skunkbush sumac |
| *Ribes pinetorum* | orange gooseberry |
| *Robinia neomexicana* | New Mexico locust |
| *Rubus arizonicus* | rubus |
| *Rumex hymenosepalus* | canaigre |
| *Sabal uresana* | Sonora palmetto |
| *Salazaria mexicana* | paper bag bush |
| *Salix* | willow |
| *Sambucus mexicana* | Mexican elderberry |
| *Sambucus vestita* | canyon elderberry |
| *Samuela carnerosana* (syn)<br>  see *Yucca carnerosa* | samandoca |
| *Sapium biloculare* | jumping bean |
| *Sapindus saponaria*<br>  var *drummondii* | soapberry |
| *Sedum stenopetalum* | stonecrop |
| *Selaginella rupicola* | selaginella |
| *Senecia neomexicanus* | New Mexico groundsel |
| *Senecio salignus* | willow groundsel |
| *Simmondsia chinensis*<br>  *Simmondsia californica* (syn) | jojoba |
| *Solanum douglasii* | Douglas nightshade |
| *Solanum elaeagnifolium* | tomatillo |
| *Solanum hindsianum* | shrub nightshade |
| *Sphaeralcea ambigua* | desert globemallow |
| *Sphaeralcea pedata* | wild hollyhock |
| *Spirostachys occidentalis* (syn)<br>  see *Allenrolfea occidentalis* | iodine bush (pickle weed) |
| *Stenolobium stans* | yellow trumpet |
| *Streptanthus arizonicus* | Arizona streptanthus |
| *Suaeda* | seepweed |
| *Symphoricarpos oreophilus* | mountain snowberry |
| *Tamarix* | tamarisk |
| *Tecoma stans* (syn)<br>  see *Stenolobium stans* | yellow trumpet |
| *Tetradymia spinosa* | horse brush |
| *Tidestromia* spp. | tidestromia |
| *Tillandsia recurvata* | tillandsia (heno pequeno, bullmoss) |

| | |
|---|---|
| *Trianthema* | horse purslane |
| *Trianthema portulacastrum* | pigweed |
| *Triodia mutica* (syn) | slim tridens |
| (obsolete) | |
| *Tridens muticus* | slim tridens |
| *Triticum* | wheat |
| *Trixis angustifolia* | trixis |
| var *angustifolia* | |
| *Trixis californica* | American trixis |
| *Tropaeoleum majus* | common nasturtium |
| *Turnera diffusa* | damiana |
| *Typha domingensis* | cattail |
| *Vaccinium scoparium* | grouse whortleberry |
| *Vauquelinia californica* | rosetree |
| *Veatchia cedrosensis* (syn) | elephant tree |
| see *Pachycormus discolor* | |
| *Verbena* spp. | verbena |
| *Vicia americana* | American vetch |
| *Vicia faba* | broadbean |
| *Viguiera deltoidea* | goldeneye |
| *Viscainoa geniculata* | viscainoa |
| *Vitis arizonica* | Arizona grape |
| *Washingtonia filifera* | fan palm |
| *Washingtonia robusta* | fan palm |
| *Welwitschia mirabilis* | welwitschia |
| *Xanthium commune* | cocklebur |
| *Yucca arborescens* (syn) | joshua tree |
| see *Yucca brevifolia* | |
| *Yucca baccata* | banana yucca |
| *Yucca brevifolia* | joshua tree |
| *Yucca carnerosana* | samandoca |
| *Yucca elata* | palmilla |
| *Yucca macrocarpa* | Schotts yucca |
| *Yucca radiosa* | palmilla |
| *Yucca schidergera* | mojave yucca |
| *Yucca valida* | datilillo |
| *Zea Mays* | maize, Indian corn |
| *Zephyranthes longiflora* | copper zephry lily |
| *Zinnia acerosa* | redstar zinnia |
| *Zizyphus* | jujube |
| *Zizyphus lycioides* | gray thorn |
| var *canescens* | |
| see *Candalia lycioides* | |
| *Zygophyllum* | zygophyllum |

# Bibliography

This bibliography includes all major publications of individuals associated with the Desert Laboratory. Also included are references with a significant relationship to the work of the Desert Laboratory.

ANONYMOUS
   1910   The plant life of the Arizona desert. [Discussion of D. T. Mac-Dougal's and V. M. Spalding's work at the Desert Botanical Laboratory.] Scottish Geographical Magazine 26(1): 9–17.

ASHBY, E.
   1931   Comparison of two methods of measuring stomatal aperture. Plant Physiology 6: 715–19.
   1932   Transpiratory organs of *Larrea tridentata* and their ecological significance. Ecology 13: 182–88.
   1933   Modern concepts of xerophytes. School Science Review 55.

AXELROD, D. I.
   1950   Studies in Late Tertiary paleobotany. *In* Evolution of Desert Vegetation in Western North America. Carnegie Institution of Washington, Publication 590.

BENSON, LYMAN
   1969   *The Cacti of Arizona.* University of Arizona Press, Tucson. 218 pp.
   ——— and R. A. DARROW
   1981   The Trees and Shrubs of the Southwestern Deserts (third edition). University of Arizona Press, Tucson. 416 pp.

BILLINGS, W. D.
   1951   Vegetational zonation in the Great Basin of western North America. International Union of Biological Sciences, Series B. 9: 101–22.

BLAKE, W. P.
   1914   The Cahuilla basin and the desert of the Colorado. *In* D. T. Mac-Dougal and Collaborators, The Salton Sea: A study of the geography, the geology, the geology, the floristics, and the ecology of a desert basin. Carnegie Institution of Washington, Publication 193:1–12.

BLUMER, J. C.
1907   A simple plan for collectors of ecological sets of plants. Plant World
       10: 40–42
1908a  Distributional features of some southwestern shrubs. Plant World
       11: 117–23.
1908b  Some effects of frost in the southwest. Torreya 8: 25–26.
1908c  Some observations on Arizona fungi. Plant World 11: 14–17.
1909a  An Arizona mesa. Plant World 12: 7–10.
1909b  Observations on cacti in cultivation. Plant World 12: 162–64.
1909c  On the plant geography of the Chiricahua Mountains. Science, n.s.
       30: 720–24.
1910a  An animal factor in plant distribution. Plant World 13: 16–18
1910b  A comparison between two mountain sides. Plant World 13: 134–
       40.
1910c  Fire as a biological factor. Plant World 13: 42–44.
1910d  Mistletoe in the southwest. Plant World 13: 240–46.
1910e  Notes on growth of pine seedlings. Plant World 13: 296–97.
1911   Change of aspect with altitude. Plant World 14:236–48.
1912a  The Euphorbias of Tucson and vicinity. Muhlenbergia 8: 97–102.
1912b  A northern prickly pear. Plant World 16: 210–212.
1912c  Notes on the phytogeography of the Arizona Desert. Plant World
       15: 183–89.
BLYDENSTEIN, J., et al.
1957   Effect of domestic livestock exclusion on vegetation in the Sonoran
       Desert. Ecology 38: 522–26.
BRANNON, M. A.
1914   The action of Salton Sea water on vegetable tissues. In D. T. Mac-
       Dougal and Collaborators, The Salton Sea: A study of the geogra-
       phy, the geology, the floristics, and the ecology of a desert basin.
       Carnegie Institution of Washington, Publication 193: 71–78.
BREAZEALE, J. F., and H. V. SMITH
1930   Caliche in Arizona. Arizona Agriculture Experiment Station Bulle-
       tin 131: 419–41.
BRIGGS, L. J., and H. L. SHANTZ
1912   The wilting coefficient and its indirect determination. Botanical
       Gazette 53: 20–37.
BROWN, J. G.
1915   The effect of dessication on the structure of Echinocactus wislizeni.
       Physiological Researches 1(6): 316–25.
1920   Subcortical formation and abnormal development of stomata in
       etiolated shoots of Opuntia blakeana. Botanical Gazette 70: 295–307.
BROWN, W. H.
1912   The relation of evaporation to the water content of the soil at the
       time of wilting. Plant World 15: 121–34.
BRYSON, R. A.
1957   The annual march of precipitation in Arizona, northern Mexico,
       and Northwestern New Mexico. University of Arizona, Tucson,
       Institute of Atmospheric Physics, Technical Report on the
       Meteorology and Climatology of Arid Regions 6. 24 pp.

BIBLIOGRAPHY

CALDWELL, J. S.
1913  The relation of environmental conditions to the phenomenon of permanent wilting in plants. Physiological Researches 1(1): 1–56.

CANNON, W. A.
1904  Observation on the germination of *Phoradendron villosum* and *P. californicum.* Torrey Botanical Club, Bulletin 31: 435–43.
1905a A new method of measuring the transpiration of plants in place. Torrey Botanical Club. Bulletin 32: 515–29.
1905b On the transpiration of *Fouquieria splendens.* Torrey Botanical Club, Bulletin 32: 397–414
1905c On the water-conducting system of some desert plants. Botanical Gazette 39: 397–408.
1906a Biological relations of certain cacti. American Naturalist 40: 27–46.
1906b The Desert Botanical Laboratory of the Carnegie Institution of Washington. Out West 24(1): 25–38.
1906c Two miles up and down in an American desert. Plant World 9: 49–55.
1907  An electric thermoregulator for paraffine baths and incubators. Plant World 10: 262–64.
1908a Acclimatization of plants at Del Monte, California. Plant World 11: 113–14.
1908b On the electric resistance of solutions of salt plants and solutions of alkali soils. Plant World 11: 10–14.
1908c Plant immigrants at Del Monte. Out West 28(5): 357–66.
1908d A redwood sport. Plant World 11: 232–34.
1908e The topography of the chlorophyll apparatus in desert plants. Carnegie Institution of Washington, Publication 98, pt. 1. 42 pp.
1909  Studies in heredity as illustrated by the trichomes of species and hybrids of *Juglans, Oenothera, Papaver,* and *Solanum.* Carnegie Institution of Washington, Publication 117. 67 pp.
1911  The root habits of desert plants. Carnegie Institution of Washington, Publication 131. 96 pp.
1912a Deciduous rootlets of desert plants. Science 35: 632–33.
1912b Some features of the root-systems of the desert plants. Popular Science Monthly 81: 90–99.
1912c Structural relations in xenoparasitism. American Naturalist 46: 675–81.
1913a Botanical features of the Algerian Sahara. Carnegie Institution of Washington, Publication 178. 81 pp.
1913b A note on a chaparral-forest relation at Carmel, California. Plant World 16: 36–38.
1913c Notes on root variation in some desert plants. Plant World 16: 323–41.
1913d Some features of the physiography and vegetation of the Algerian Sahara. American Geographical Society, Bulletin 45(7): 481–89.
1913e Some relations between root characters, ground water and species distribution. Science, n. s. 37(950): 420–23.

1913*f* Some relations between salt plants and salt-spots. *In* Dudley Memorial Volume, pp. 123–29. Stanford University Publications, University Series 11.

1914*a* A note on the reversibility of the water reaction in a desert liverwort. Plant World 17: 261–65.

1914*b* On the density of the cell sap in some desert plants. Plant World 17: 209–12.

1914*c* Recent exploration in the western Sahara. American Geographical Society, Bulletin 46(2): 81–99.

1914*d* Specialization in vegetation and in environment in California. Plant World 17: 223–37.

1914*e* Tree distribution in Central California. Popular Science Monthly 85: 417–24.

1915   On the relation of root growth and development to the temperature and aeration of the soil. American Journal of Botany 2: 211–24.

1916   Distribution of the cacti with especial reference to the role played by the root response to soil temperature and soil moisture. American Naturalist 50: 435–42.

1917   Relation of the rate of root growth in seedlings of *Prosopis velutina* to the temperature of the soil. Plant World 20: 320–33.

1918*a* The evaluation of the soil temperature factor in root growth. Plant World 21: 64–67.

1918*b* Some biological features of roots. Scientific American 119: 373.

1920   The ecological relations of roots. Science n.s. 51: 393–94.

1921   Plant habits and habitats in the arid portions of South Australia. Carnegie Institution of Washington, Publication 308, 139 pp.

1923   The influence of the temperature of the soil on the relation of roots to oxygen. Science, n.s. 58:331–32.

1924*a* General and physiological features of the vegetation of the more arid portions of southern Africa, with notes on the climatic environment. Carnegie Institution of Washington, Publication 354. 159 pp.

1924*b* A note on the relation of root growth in the soil to the oxygen supply: the growth ratio. Ecology 319–21.

1925*a* On the upper critical concentration of oxygen in root growth. Science, n.s. 61:118–20.

1925*b* Physiological features of roots with especial references to the relation of roots to the aeration of the soil. Carnegie Institution of Washington, Publication 368. 168 pp.

————, and E. E. FREE

1917   The ecological significance of soil aeration. Science, n.s. 45: 178–80.

CLARKE, B. L.

1925   Studies in swelling. I: The swelling of agar-agar gels as a function of water content before swelling. American Chemical Society, Journal 47: 1954–58.

CLEMENTS, F. E.

1936   The origin of the desert climax and climate. *In* T. H. Goodspeed, Ed., Essays in Geobotany in Honour of William Albert Setchell, pp. 87–140. University of California Press, Berkeley.

COOKE, R. U.
1970    Stone pavements in deserts. Association of American Geographers, Annals 60(3): 560–77.
COOPER, W. S.
1922    The broad-sclerophyll vegetation of California; an ecological study of the chaparral and its related communities. Carnegie Institution of Washington Publication 319. 124 pp.
Coville, F. V.
1893    Botany of the Death Valley expedition. Contributions from the U.S. National Herbarium. Vol. IV. 363 pp.
1904    Desert plants as a source of drinking water. Smithsonian Institution, Annual Report 1903: 499–505.
COVILLE, F. V., and D. T. MACDOUGAL
1907    Desert Botanical Laboratory of the Carnegie Institution. Carnegie Institution of Washington, Publication 6. 58 pp.
DARROW, R. A.
1935    A study of the transpiration rates of several desert grasses and shrubs as related to environmental conditions and stomatal periodicity. University of Arizona M.S. Thesis. 144 pp.
1943    Vegetative and floral growth of *Fouquieria splendens*. Ecology 24(3): 30–32.
DAUBENMIRE, R. F.
1974    Plants and environment. A textbook of plant autecology (third edition). John Wiley & Sons, New York, 422 pp.
DICE, L. R.
1939    The Sonoran biotic province. Ecology 20: 118–29.
————, and P. M. BLOSSOM
1937    Studies of mammalian ecology in southwestern North America, with special attention to the colors of desert mammals. Carnegie Institution of Washington Publication 485. 129 pp.
FRAPS, M.
1930    Studies on respiration and glycolysis in *Planaria*. I: Methods and certain basic factors in respiration. II: Basic environmental factors and oxygen consumption of *Planaria* sp. Physiological Zoology 3(2): 242–70.
FREE, E. E.
1911    Studies in soil physics. Plant World 14: 29–39, 59–66, 110–19, 164–76, 186–90.
1914    Sketch of the geology and soils of the Cahuilla Basin. *In* D. T. MacDougal and Collaborators, The Salton Sea; A study of the geography, the geology, the floristics, and the ecology of a desert basin. Carnegie Institution of Washington, Publication 193: 21–33.
1917    Note on the swelling of gelatine and agar gels in solutions of sucrose and dextrose. Science, n. s. 46: 142–43.
1918    A colloidal hypothesis of protoplasmic permeability. Plant World. 21: 141–50.
GENTRY, H. S.
1942    Rio Mayo plants; a study of the flora and vegetation of the valley of the Rio Mayo, Sonora. Carnegie Institution of Washington, Publication 527. 328 pp.

1957  Los Pastizales de Durango, estudio ecológico, fisiográfico y florís-
tico. Instituto Mexicano de Recursos Naturales Renovables, A. C.
Mexico, D. F.

HALL, H. M., and F. L. LONG
1921  Rubber content of North American Plants. Carnegie Institution of
Washington Publication 313.

HARRIS, J. A.
1916  The variable desert. Scientific Monthly 3: 41–50.

————., J. V. LAWRENCE, and R. A. GORTNER
1915  On the osmotic pressure of the juices of desert plants. Science, n.s.
41: 656–58.

————., and J. V. LAWRENCE, with the cooperation of R. A. GORTNER
1916  The cryoscopic constants of expressed vegetable saps as related to
local environmental conditions in the Arizona desert. Physiological
Researches 2(1): 1–49.

————., R. A. GORTNER, W. F. HOFFMANN, and A. T. VALENTINE
1924  The osmotic concentration, specific electrical conductivity, and
chloride content of the tissue fluids of the indicator plants of
Tooele Valley, Utah. Journal of Agricultural Research 27:893–
924.

HASKINS, C. P.
1967  The search for understanding. Carnegie Institution of Washington.
330 pp.

HASSE, E. F.
1970  Environmental fluctuations on south-facing slopes in the Santa
Catalina Mountains of Arizona. Ecology 51(6): 959–74.

HASTINGS, J. R.
1959  Vegetation change and arroyo cutting in southeastern Arizona.
Arizona Academy of Science, Journal 1: 60–67.
1961  Precipitation and saguaro growth. University of Arizona Arid Lands
Colloquia, 1959/60/61. pp. 30–38.
1963  Historical changes in the vegetation of a desert region. University
of Arizona (Ph.D. Dissertation). 499 pp.

HASTINGS, J. R., ed.
1964  Climatological data for Baja California. University of Arizona, Institute of Atmospheric Physics, Technical Reports on the
Meteorology and Climatology of Arid Regions 14. 132 pp.
1964  Climatological data for Sonora and Northern Sinaloa. University of
Arizona Institute of Atmospheric Physics, Technical Reports on the
Meteorology and Climatology of Arid Regions 15. 152 pp.

———— and S. M. ALCORN
1961  Physical determination of growth and age in the giant cactus.
Arizona Academy of Science, Journal 2 (1):32–39.

———— and R. R. HUMPHREY, eds.
1969  Climatological data and statistics for Baja California. University of
Arizona Institute of Atmospheric Physics, Technical Reports on the
Meteorology and Climatology of Arid Regions 18. 95 pp.
1969  Climatological data and statistics for Sonora and Northern Sinaloa.
University of Arizona, Technical Reports on the Meteorology and
Climatology of Arid Regions 19. 96 pp.

HASTINGS, J. R., *continued*
——— and R. M. TURNER
1965  The changing mile; an ecological study of vegetation changes with time in the lower mile of an arid and semi-arid region. University of Arizona Press, Tucson. 317 pp.
———, R. M. TURNER, and D. K. WARREN
1972  An atlas of some plant distributions in the Sonoran Desert. University of Arizona, Institute of Atmospheric Physics, Technical Reports on the Meteorology and Climatology of Arid Regions 21. 255 pp.
HORNADAY, W. T.
1908  Camp-fires on desert and lava. Scribner's Sons and Co., New York. 366 pp.
HOWARD, W. L.
1946  Luther Burbank, Chronica Botanica 9: 304–498.
HUMPHREY, R. R.
1931  Thorn formation in *Fouquieria splendens* and *Idria columnaris.* Torrey Botanical Club, Bulletin 58: 263–64.
1932  The morphology, physiology and ecology of *Coldenia canescens.* Ecology 13: 153–58.
1933  A detailed study of desert rainfall. Ecology 14: 31–34.
1935  A study of *Idria columnaris* and *Fouquieria splendens.* American Journal of Botany 22: 184–207.
1963  Growth habits of barrel cactus. Madroño 3: 448–52.
1970  Five dominants of the central desert of Baja California. Cactus and Succulent Journal 42(5): 207–15.
1974  The Boojum and its home. University of Arizona Press, Tucson. 214 pp.
HUNTINGTON, E.
1911  The greenest of deserts. Harper's Magazine 123(733): 50–58.
———, with contributions by C. SCHUCHERT, A. E. DOUGLASS, and C. J. KULLMER
1914  The climatic factor, as illustrated in arid America. Carnegie Institution of Washington Publication 192. 341 pp.
IVES, R. L.
1949  Climate of the Sonoran Desert Region. Association of American Geographers 39(3): 43–187.
1964  The Pinacate Region, Sonora, Mexico. California Academy of Sciences, Occasional Papers 47. 43 pp.
JOHNSON, E. S., and B. E. LIVINGSTON
1916  Measurement of evaporation rates for short term intervals. Plant World 19: 136–40.
JOHNSON, D. S.
1918  The fruit of *Opuntia fulgida;* a study of perennation and proliferation in the fruits of certain Cactacea. Carnegie Institution of Washington, Publication 269. 62 pp.
JOHNSTON, I. M.
1941  Gypsophily among Mexican desert plants. Arnold Arboretum, Journal 22: 145–70.

1943–44   Plants of Coahuila, eastern Chihuahua, and adjoining Zacatecas and Durango. 5 parts. Arnold Arboretum, Journal 24: 306–39, 375–421; 25: 43–83, 431–53, 133, 182.

JONES, J. C.
1914   The tufa deposits of the Salton Sink. *In* D. T. MacDougal and Collaborators, The Salton Sea: A study of the geography, the geology, the floristics, and the ecology of a desert basin. Carnegie Institution of Washington, Publication 193. 79–84.

KIRKWOOD, J. E.
1909   Some Mexican fiber plants. Plant World 12: 25–34.

LIVINGSTON, B. E.
1906a  Note on the relation between growth of roots and tops in wheat. Botanical Gazette 41: 139–43. (Contributions from the Hull Botanical Laboratory. LXXXI.)

1906b  Paraffined wire pots for soil cultures. Plant World 9: 62–66.

1906c  The relation of desert plants to soil moisture and to evaporation. Carnegie Institution of Washington, Publication 50. 78 pp.

1906d  A simple method for experiments with water cultures. Plant World 9: 13–16.

1907a  Evaporation and plant development. Plant World 10: 269–76.

1907b  Relative transpiration in cacti. Plant World 10: 110–14.

1908a  The botanical garden at Pisa. Plant World 11: 156–57.

1908b  The botanical garden of Florence. Plant World 11: 106–12.

1908c  Evaporation and centers of plant distribution. Plant World 11: 106–12.

1908d  Evaporation and plant habitats. Plant World 11: 1–9.

1908e  A method for controlling plant moisture. Plant World 11: 39–40.

1908f  A new method for cultures of algae and mosses. Plant World 11: 183–84.

1908g  A simple atmometer. Science 28: 319–20.

1909a  The heath of Lueneburg. Plant World 12: 231–37.

1909b  Present problems of physiological plant ecology. American Naturalist 43: 369–78. Also *in* Plant World 12: 41–46.

1909c  A repeated cycle in assimilation. Plant World 12: 66–67.

1909d  Roles of the soil in limiting plant activities. Plant World 12: 49–53.

1909e  Soils of the Desert Laboratory domain. *In* V. M. Spalding, Distribution and movements of desert plants. Carnegie Institution of Washington, Publication 113: 83–94.

1909f  Stomata and transpiration in *Tradescantia zebrina.* Science, n.s. 29: 269–70.

1910a  Evaporation as a climatic factor influencing vegetation. Horticultural Society of New York, Memoir 2: 43–54.

1910b  Operation of the porous cup atmometer. Plant World 13: 111–19.

1910c  A rain-correcting atmometer for ecological instrumentation. Plant World 13: 79–82.

1910d  Relation of soil moisture to desert vegetation. Botanical Gazette 50:241–56.

1911a  Light intensity and transpiration. Botanical Gazette 52: 417–38.

LIVINGSTON, B. E., *continued*

1911*b* Paper atmometers for studies in evaporation and plant transpiration. Plant World 14: 281–89.

1911*c* A radio-atmometer for comparing light intensities. Plant World 14: 96–99.

1911*d* A study of the relation between summer evaporation intensity and centers of plant distribution in the United States. Plant World 14: 205–22.

1912*a* The choosing of a problem for research in plant physiology. Plant World 15: 73–82.

1912*b* Incipient drying in plants. Science, n.s. 35: 394–95. (Abstract)

1912*c* Present problems in soil physics as related to plant activities. American Naturalist 46: 294–301.

1912*d* A rotating table for standardizing porous cup atmometers. Plant World 15: 157–62.

1912*e* A schematic representation of the water relations of plants, a pedagogical suggestion. Plant World 15: 214–18.

1913*a* Adaptation in the living and non-living. American Naturalist 47: 72–82.

1913*b* Climatic areas of the United States as related to plant growth. American Philosophical Society, Proceedings 52: 257–75.

1913*c* IV. Osmotic pressure and related forces as environmental factors. Plant World 16: 165–76.

1913*d* The resistance offered by leaves to transpirational water loss. Plant World 16: 1–35.

1915*a* Atmometry and the porous cup atmometer. Plant World I 18: 21–30, II 18: 51–74, III 18: 95–111, III 18: 143–49.

1915*b* Atmospheric influence on evaporation and its direct measurement. Monthly Weather Review 43(3): 126–131.

—— and W. H. BROWN

1912    Relation of the daily march of transpiration to variations in the water content of foliage leaves. Botanical Gazette 53: 309–30.

—— and A. H. ESTABROOK

1912    Observations on the degree of stomatal movement in certain plants. Torrey Botanical Club, Bulletin 39: 15–22.

—— and L. A. HAWKINS

1915    The water-relation between plant and soil. Carnegie Institution of Washington, Publication 204: 3–48.

—— and G. J. LIVINGSTON

1913    Temperature coefficients in plant geography and climatology. Botanical Gazette 56: 349–75.

—— and E. B. SHREVE

1916    Improvements in the method for determining the transpiring power of plant surfaces by hygrometric paper. Plant World 19(10): 287–309.

—— and F. SHREVE

1921    Distribution of vegetation in the United States as related to climatic conditions. Carnegie Institution of Washington, Publication 284. 590 pp.

LIVINGSTON, G. J.
1908   An annotated bibliography of evaporation. Monthly Weather Review 36: 181–86, 301–6, 375–81.
1909   An annotated bibliography of evaporation. Monthly Weather Review 37: 68–72, 103–9, 157–60, 193–99, 248–52.

LLOYD, F. E.
1904   A visit to the Desert Botanical Laboratory. New York Botanical Garden, Journal 5: 172–77.
1905a  The artificial induction of leaf formation in the ocotillo. Torreya 5: 175–79. Also in Plant World 9: 56–62.
1905b  A botanical laboratory in the desert. Popular Science Monthly 66: 329–42.
1905c  The Desert Botanical Laboratory of the Carnegie Institution of Washington. Biologisches Centralblatt.
1905d  Isolation and the origin of species. Science, n.s. 22: 710–12.
1906   Paloverde: the evergreen tree of the desert. Plant World 9: 165–71.
1907a  Observations of the flowering periods of certain cacti. Plant World 10: 31–39.
1907b  Pima Cañón and Castle Rock in the Santa Catalina Mountains. Plant World 10: 251–59.
1908a  Extra-floral nectaries in the cacti. Plant World 11: 138–40.
1908b  Methods of vegetative reproduction in guayule and mariola. Plant World 11: 201–8.
1908c  A perennial dodder. Plant World 11: 40–41.
1908d  The physiology of stomata. Carnegie Institution of Washington, Publication 82. 142 pp.
1908e  Some features of the anatomy of guayule (*Parthenium argentatum* Gray). Plant World 11: 172–79.
1908f  Some seedlings of the desert. Plant World 11: 154–56.
1908g  A water-storage organ in Cuscuta. Plant World 11: 67–68.
1909   Overlapping habitats. Plant World 12: 73–78.
1911   The behavior of tannin in persimmons, with some notes on ripening. Plant World 14: 1–14.
1911b  Guayule (*Parthenium argentatum* Gray), a rubber-plant of the Chihuahuan Desert. Carnegie Institution of Washington, Publication 139. 213 pp.
1912   The relation of transpiration and stomatal movements to the water-content of the leaves in *Fouquieria splendens*. Plant World 15: 1–14.
1917a  The analogy between the behavior of gelatine and that of protoplasm. Royal Society of Canada, Transactions.
———— and C. S. RIDGWAY
1912   The behavior of the nectar gland in the cacti, with a note on the development of the trichomes and areolar cork. Plant World 15: 145–56.
———— and V. ULEHLA
1926   The role of the wall in the living cell as studied by auxographic method. Royal Society of Canada, Transactions 20: 45–73.

LONG, E. R.
1915a Acid accumulation and destruction in large succulents. Plant World 18: 261–72.
1915b Chemical changes accompanying desiccation and partial starvation. Physiological Researches 1(6): 298–315.
1915c Growth and colloid hydration in cacti. Botanical Gazette 59: 491–97.
1918 Further results in desiccation and respiration of *Echinocactus*. Botanical Gazette 65: 354–58.

LONG, F. L.
1919 The quantitative determination of photosynthetic activity in plants. Physiological Researches 2(6): 277–300.

MABRY, T. J., DR. DIFEO JR. eds.
1977 Creosote bush: biology and chemistry of *Larrea* in new world deserts. US/IBP Synthesis Series 6. Dowden, Hutchinson and Ross, Stroudsburg, Penn. 284 pp.

MACDOUGAL, D. T.
1903a Some aspects of desert vegetation. Plant World 6: 249–37.
1903b Soil Temperature and vegetation. Monthly Weather Review 31: 375.
1904a Botanical explorations in the southwest. New York Botanical Gardens, Journal 5: 89–91.
1904b Delta and desert vegetation. Botanical Gazette 38: 44–63.
1905a The delta of the American Nile. American Geographical Society, Bulletin 37.
1905b Discontinuous variation and the origin of species. Torreya 5: 1–6. Also *in* Science, n.s. 21: 540–43.
1905c Studies in organic evolution. New York Botanical Garden, Journal 6: 27–36.
1905d Suwarro or saguaro. New York Botanical Garden, Journal 6: 149–150.
1905e The suwarro, or tree cactus. New York Botanical Garden, Journal 6: 129–133.
1906a The delta of the Rio Colorado. American Geographical Society, Bulletin 38(1): 1–16.
1906b Discontinuous variation in pedigree-cultures. Popular Science Monthly 69: 207–25.
1906c [Discussion of] Discontinuous variation [see MacDougal (1905b) above], Science, n.s. 24: 730–31.
1906d Heredity, and the origin of species. Monist 16: 32–64. Also *in* New York Botanical Garden, Contribution 80. 32 pp. (printed 1905, in advance of the Monist). Also *in* Smithsonian Institution, Annual Report, 1908, p. 505–23.
1907a Collecting Cacti in southern Mexico. New York Botanical Garden, Journal 8: 1–13.
1907b The desert basins of the Colorado Delta. American Geographical Society, Bulletin 39(12): 705–29.
1907c Factors affecting the seasonal activities of plants. Plant World 10: 217–37.
1907d The faculties of plants. Scientific American, Supplement.

1907*e* Hybridization of the oaks. Scientific American, Supplement. 63: 26,105–26,106.

1907*f* Hybridization of wild plants. Botanical Gazette 43: 45–58.

1907*g* Hybrids among wild plants. Plant World 10: 25–27.

1907*h* Natural hybrids. Plant World 10: 138–139.

1907*i* A realized mirage: Salton Sea, the wonderful desert lake where wood sinks and stones float, explored in a sailboat. Discovery 1: 1–4.

1907*j* A voyage below sea-level on the Salton Sea. Outing 51: 592–602.

1908*a* Across Papagueria. Plant World 11: 93–99, 123–131. Also *in* American Geographical Society, Bulletin 40. 21 pp.

1908*b* Botanical features of North American deserts. Carnegie Institution of Washington, Publication 99. 112 pp.

1908*c* The course of the vegetative seasons in southern Arizona. Plant World 11: 189–201, 217–31, 237–49,,261–70.

1908*d* Heredity and environic forces. Science, n.s. 27: 123–28.

1908*e* The physiological aspect of a species. American Naturalist 42: 249–52.

1908*f* The poisonous action of loco-weed and ladies' slipper. Plant World 11: 85–86.

1908*g* Problems of the desert. Plant World 11: 28–39.

1908*h* The vegetation of the Tucson region. University of Arizona Monthly 9(7): 1–18.

1909*a* Darwinism and experimentation in botany. Plant World 12: 97–101, 121–27.

1909*b* The direct influence of environment. p. 114–42. *In* American Association for the Advancement of Science, Fifty years of Darwinism; modern aspects of evolution. Holt and Co., New York.

1909*c* Influence of aridity upon the evolutionary development of plants. Plant World 12: 217–31.

1909*d* The origin of desert floras. *In* V. M. Spalding, Distribution and movements of desert plants. Carnegie Institution of Washington, Publication 113: 113–19.

1909*e* Origination of self-generating matter and the influence of aridity upon its evolutionary development. Journal of Geology 17: 603–22. Same *in* MacDougal (1910*b*)

1910*a* The making of parasites. Plant World 13: 207–14.

1910*b* Origination of self-generating matter, and the influence of aridity upon its evolutionary development, p. 278–97. *In* W. Bailey and R. D. Salisbury, eds., Outlines of geologic history with especial reference to North America. University of Chicago Press, Chicago.

1910*c* Plant parasites. Scientific American, Supplement 70(1800): 6–8.

1911*a* Alterations in heredity induced by ovarial treatments. Botanical Gazette 51: 241–57.

1911*b* An attempted analysis of parasitism. Botanical Gazette 52: 249–60.

1911*c* Climatic selection in a hybrid progeny. Plant World 14: 129–31.

1911*d* Environment and heredity, modern studies in experimental evolution. Scientific American, Supplement 71: 122–23.

MacDougal, D. T., *continued*
1911*e* The inheritance of habitat effects by plants. Plant World 14: 53–59.
1911*f* Induced and occasional parasitism. Torrey Botanical Club, Bulletin 38: 473–79.
1911*g* Organic response. American Naturalist 45: 5–40.
1911*i* The water relations of desert plants. Popular Science Monthly 79: 540–53.
1912*a* North American deserts. Geographical Journal 39(2): 105–23.
1912*b* Some physical and biological features of North American deserts. Scottish Geographical Magazine 28(9): 499–56.
1912*c* The water-balance of desert plants. Annals of Botany 26: 71–93.
1913*a* Aus Nordamerikas wuesten. Die Erde 1. 338 pp.
1913*b* By caravan through the Libyan Desert. Harper's Magazine 127(760): 489–500.
1913*c* The deserts of western Egypt. Plant World 16: 291–303.
1913*d* The determinative action of environic factors upon *Neobeckia aquatica* Greene. Flora 106: 264–80.
1913*e* From the Red Sea to the Nile. Plant World 16: 243–55.
1914*a* The auxo-thermal integration of climatic complexes. American Journal of Botany 1: 186–93.
1914*b* The effect of potassium iodide, methylene blue and other substances applied to the embryo sacs of seed-plants. Society for Experimental Biology and Medicine. Proceedings 12: 1–3.
1914*c* The measurement of environic factors and their biologic effects. Popular Science Monthly 84: 417–33.
1914*d* Movements of vegetation due to submersion and desiccation of land areas in the Salton Sink. *In* D. T. MacDougal and Collaborators, The Salton Sea: A study of the geography, the geology, the floristics, and the ecology of a desert basin. Carnegie Institution of Washington, Publication 193: 115–72.
1915*a* General course of depletion in starving succulents. Physiological Researches 1(6): 292–98.
1915*b* Light and rate of growth in plants. Science, n.s. 41: 467–69.
1915*c* The Salton Sea. American Journal of Science 39: 231–50.
1916*a* Biological research institutions; organizations, men, and methods. Scripps Institute of Biological Research, Bulletin 1.
1916*b* Imbibitional swelling of plants and colloidal mixtures. Science, n.s. 44(1136): 502–5.
1916*c* The mechanism and conditions of growth. New York Botanical Garden, Memoirs 6: 5–26.
1917*a* The beginnings and physical basis of parasitism. Plant World 20: 238–44.
1917*b* A decade of the Salton Sea. Geographical Review 3(6): 457–73.
1917*c* The relation of growth and swelling of plants and biocolloids to temperature. Society for Experimental Biology and Medicine. Proceedings 15: 48–50.
1917*d* To increase the yield of food. North American Review 206(740): 62–69.

1918*a* Effect of bog and swamp waters on swelling in plants and in biocolloids. Plant World 21: 88–99.

1918*b* The trend of research in evolution and the utilization of its concepts. University of California, School for Subtropical Horticulture, Bulletin.

1919*a* Growth in organisms. Science, n.s. 49: 599–605.

1919*b* Hydration and growth. American Philosophical Society, Proceedings 58: 346–72.

1920*a* Auxographic measurement of swelling of biocolloids and of plants. Botanical Gazette 70: 126–36.

1920*b* Hydration and growth. Carnegie Institution of Washington, Publication 297. 176 pp.

1920*c* The physical factors in the growth of the tomato. Torrey Botanical Club, Bulletin 47: 261–69.

1921*a* The action of bases and salts on biocolloids and cell-masses. American Philosophical Society, Proceedings 60: 15–30.

1921*b* The distentive agencies in the growth of the cell. Society for Experimental Biology and Medicine, Proceedings 19: 103–10.

1921*c* Effects of age and of the inclusion of salts on the heterotropic action of colloidal bodies of cytological interest. Society for Experimental Biology and Medicine, Proceedings 18: 244–46.

1921*d* Growth in trees. Carnegie Institution of Washington, Publication 307. 41 pp.

1921*e* Growth in trees. [A summary of MacDougal (1921*d*), above.] American Philosophical Society, Proceedings 60: 7–14.

1921*f* How mountain plants behave when they go to the seaside. Garden 34: 305.

1921*g* A new high temperature record for growth. Science, n.s. 53: 370–72.

1921*h* The reactions of plants to new habitats. Ecology 2: 1–20.

1921*i* Water deficit and the action of vitamines, amino-compounds, and salts on hydration. American Journal of Botany. 8: 296–302.

1922*a* The autograph of the Arizona ash tree. Garden 35: 313.

1922*b* The probable action of lipoids in growth. American Philosophical Society, Proceedings 61: 33–52.

1922*c* Sugar and spines. Garden 35: 249–51.

1923    Permeability and the increase in volume of contents of living and of artificial cells. American Philosophical Society, Proceedings 62: 1–25.

1924*a* The arrangement and action of material in the plasmatic layers and cell-walls of plants. American Philosophical Society, Proceedings 63: 76–93.

1924*b* Dendrographic measurements. *In* Growth in trees and massive organs of plants. Carnegie Institution of Washington, Publication 350: 1–88.

1925*a* Absorption and exudation pressures of sap in plants. American Philosophical Society, Proceedings 64: 102–30.

1925*b* Accretion and distention in plant cells. American Naturalist 59: 336–45.

MacDougal, D. T., *continued*
1925*c* A cycle of the Salton Sea. *In* Festschrift Carl Schröter. Geobotanisches Institut Rübel, Zürich, Veröffentlichungen 3: 345–63.
1925*d* Growth in trees. Scientific Monthly 21: 99–103.
1925*e* Reversible variations in volume, pressure, and movements of sap in trees. Carnegie Institution of Washington, Publication 365. 90 pp.
1925*f* Tree trunks, growth and reversible variations in circumference. Science, n.s. 61: 370–72.
1926*a* The autobiography of trees. How trees are being made to write their own diaries. American Forests and Forest Life 32: 661–62.
1926*b* Growth and permeability of century-old cells. American Naturalist 60: 393–415.
1926*c* The hydrostatic system of trees. Carnegie Institution of Washington, Publication 373. 125 pp.
1926*d* The hydrostatic system of trees. [A summary of MacDougal (1926 *c*) above.] Scientific American 134: 378–79.
1927 Sunburn in the dark. Scientific American 137: 25–27.
1928*a* Activities in plant physiology. Scientific Monthly 26: 464–67.
1928*b* Can we grow our own rubber? Guayule, a native American rubber, producing shrub, is being cultivated on a large scale in California. Scientific American 139: 16–19.
1928*c* Substances regulating the passage of material into and out of plant cells: the lipoids. American Philosophical Society, Proceedings 67: 33–45.
1930*a* The green leaf; and major activities of plants in sunlight. D. Appleton and Co., New York, London. 141 pp.
1930*b* Lengthened growth periods and continuous growth. American Philosophical Society, Proceedings 69: 329–45.
1936 Studies in tree growth. Carnegie Institution of Washington, Publication 462.
——— and W. A. Cannon
1910 The condition of parasitism in plants. Carnegie Institution of Washington, Publication 129. 60 pp.
——— and B. L. Clarke
1925 The hydrophilic effect of ions on agar and protoplasmic components. Science, n.s. 62: 136–37.
——— and F. L. Long
1927 Characters of cells attaining great age. American Naturalist 61: 385–406.
———, E. R. Long, and J. G. Brown
1915 The end results of desiccation and starvation of succulent plants. Physiological Researches 1(6): 289–292.
——— and V. Moravek
1927 Activities of a constructed colloidal cell. Protoplasma 2: 161–88.
——— and J. B. Overton
1927 Sap flow and pressure in trees. Science, n.s. 65: 189–90.

——, J. B. OVERTON, and G. M. SMITH
1929  The hydrostatic-pneumatic system of certain trees: movements of liquids and gases. Carnegie Institution of Washington, Publication 397. 99 pp.
——, H. M. RICHARDS, and H. A. SPOEHR
1919  Basis of succulence in plants. Botanical Gazette 67: 405–16.
—— and F. SHREVE
1924  Growth in trees and massive organs. Carnegie Institution of Washington, Publication 350.
1926  Can the primitive forest ever return? National Parks Bulletin 49: 11–12.
—— and G. M. SMITH
1927  Long-lived cells of the redwood. Science 66: 456–57.
—— and E. S. SPALDING
1910  The water balance of the succulent plants. Carnegie Institution of Washington, Publication 141. 77 pp.
—— and H. A. SPOEHR
1917a The behavior of certain gels useful in the interpretation of the action of plants. Science, n.s. 45(1168): 484–88.
1917b The effects of acids and salts on "bio-colloids." Science, n.s. 46(1185): 269–72.
1917c Growth and inhibition. American Philosophical Society, Proceedings 56: 289–352.
1917d The measurement of light in some of its more important physiological aspects. Science, n.s. 45(1172): 616–18.
1918a The effect of organic acids and their amino compounds on the hydration of agar and on a biocolloid. Society for Experimental Biology and Medicine, Proceedings. 16: 33–35.
1918b The organization of xerophytism. Plant World 21: 245–49.
1919a Hydration effects of amino compounds. Society for Experimental Biology and Medicine, Proceedings 17: 33–36.
1919b The solution and fixation accompanying swelling and drying of biocolloids and plant tissues. Plant World 22: 129–37.
1920a The components and colloidal behavior of plant protoplasm. American Philosophical Society, Proceedings 59: 150–170.
1920b Swelling of agar in solutions of amino acids and related compounds. Botanical Gazette 70: 268–78.
—— and G. SYKES
1915  The travertine record of Blake Sea. Science, n.s. 42: 133–34.
——, A. M. VAIL, and G. H. SHULL
1907  Mutations, variations, and relationships of the Oenotheras. Carnegie Institution of Washington, Publication 81. 92 p. (Papers of the Station of Experimental Evolution, 9)
——, A. M. VAIL, G. H. SHULL, and J. K. SMALL
1905  Mutants and hybrids of the Oenotheras. Carnegie Institution of Washington, Publication 24. 57p. (Paper of the Station for Experimental Evolution, Cold Spring Harbor, N.Y. 1)

MacDougal, D. T., *continued*
—— and E. B. Working
1921    Another high-temperature record for growth and endurance. Science n.s. 54: 152–53.
1933    The pneumatic system of plants. Carnegie Institution of Washington, Publication 441.
—— and collaborators
1914    The Salton Sea; a study of the geography, the geology, the floristics, and the ecology of a desert basin. Carnegie Institution of Washington, Publication 193. 182 pp.
McGinnies, W. G.
1955    A report on the ecology of the arid and semi-arid areas of the United States and Canada. *In* Plant Ecology, Reviews of Research. UNESCO, Paris. Arid Zone Research 6: 250–301.
——, and J. F. Arnold
1939    Relative water requirements of Arizona range plants. Arizona Agriculture Experiment Station, Technical Bulletin 80: 165–246.
McGinnies, W. G., B. Goldman, and P. Paylore, eds.
1968    Deserts of the World: an appraisal of research into their physical and biological environments. University of Arizona Press, Tucson. 788 pp.
Mallery, T. D.
1934    Comparison of the heating and freezing methods of killing plant material for cryoscopic determinations. Plant Physiology 9: 369–75.
1935    Changes in the osmotic value of the expressed sap of leaves and small twigs of *Larrea tridentata* as influenced by environmental conditions. Ecological Monographs 5: 1–35.
1936a   Rainfall records for the Sonoran Desert, I. Ecology 17: 110–21.
1936b   Rainfall records for the Sonoran Desert, II. Summary of readings to December, 1935. Ecology 17: 212–15.
Martin, E.
1943    Studies of evaporation and transpiration under controlled conditions. Carnegie Institution of Washington Publication 550.
Martin, W. P., and J. E. Fletcher
1943    Vertical zonation of great soil groups on Mount Graham, Arizona, as correlated with climate, vegetation, and profile characteristics. Arizona Agricultural Experiment Station, Technical Bulletin 99. 153 pp.
Merriam, R.
1969    Source of sand dunes of southeastern California and northwestern Sonora, Mexico. Geological Society of America, Bulletin 80(3): 531–34.
Muller, C. H.
1938    Relations of the vegetation and climatic types in Nuevo León, México. American Midland Naturalist 21: 687–729.
1947    Vegetation and climate of Coahuila. Moderno 9: 33–57.
Nichol, A. A.
1952    The natural vegetation of Arizona. Arizona Agricultural Experiment Station, Technical Bulletin 127: 189–230.

PARISH, S. B.
   1913   Plants introduced into a desert valley as a result of irrigation. Plant
          World 16: 275–80.
   1914   Plant ecology and floristics of Salton Sink. *In* D. T. MacDougal and
          Collaborators; the Salton Sea; a study of the geography, the
          geology, the floristics, and the ecology of a desert basin. Carnegie
          Institution of Washington, Publication 193: 85–114.
   1930   Vegetation of the Mojave and Colorado Deserts of southern Cali-
          fornia. Ecology 11(4): 281–86.
PEIRCE, G. J.
   1914   The behavior of certain micro-organisms in brine. *In* D. T. Mac-
          Dougal and Collaborators, The Salton Sea: A study of the Geogra-
          phy, the geology, the floristics, and the ecology of a desert basin.
          Carnegie Institution of Washington, Publication 193: 49–70.
PHILLIPS, W. S.
   1963   Depth of roots in soil. Ecology 44(2): 424.
PULLING, H. E., and B. E. LIVINGSTON
   1915   The water-supplying power of the soil as indicated by osmometers
          Carnegie Institution of Washington, Publication 204: 49–84.
RICHARDS, H. M.
   1915   Acidity and gas interchange in cacti. Carnegie Institution of Wash-
          ington, Publication 209. 107 pp.
   1918   Determination of acidity in plant tissues. Torrey Botanical Club,
          Memoirs 17: 241–245.
ROSS, W. H.
   1914   Chemical composition of the water of Salton Sea and its annual
          variation in concentration, 1906–1911. *In* D. T. MacDougal and
          Collaborators, The Salton Sea: A study of the geography, the
          geology, the floristics, and the ecology of a desert basin. Carnegie
          Institution of Washington, Publication 193: 35–46.
RUNYON, E. H.
   1934   The organization of the creosote bush with respect to drought.
          Ecology 15: 128–38.
   1936   Ratio of water content to dry weight in leaves of the creosote bush.
          Botanical Gazette 97: 518–53.
SCHRATZ, E.
   1931a  Vergleichende untersuchungen über den wasserhaushalt von
          pflanzen im trockengebiete des südlichen Arizona. Jahrbuerker fur
          Wissenschaftliche Botanik 74(2): 153–290.
   1931b  Zum vergleich der transpiration xeromorpher und mesomorpher
          pflanzen. Journal of Ecology 19: 292–96.
SHIVE, J. W.
   1915   An improved non-absorbing porous cup atmometer. Plant World
          18: 7–10.
SHIVE, J. W., and B. E. LIVINGSTON
   1914   The relation of atmospheric evaporating power to soil moisture
          content at permanent wilting in plants. Plant World 17: 81–121.
SHREVE, E. B.
   1912   A calorimetric method for the determination of leaf temperatures.
          Johns Hopkins University Circular 1912: 146–48.

SHREVE, E. B., *continued*
1914  The daily march of transpiration in a desert perennial. Carnegie Institution of Washington, Publication 194. 94 pp.
1915  An investigation of the causes of autonomic movements in succulent plants. Plant World 18: 297–312, 331–43.
1916  An analysis of the causes of variations in the transpiring power of cacti. Physiological Researches 2(3): 73–127.
1918  Investigations on the imbibition of water by gelatine. Science, n.s. 48: 324–27.
1919a  Investigations on the absorption of water by gelatine. Franklin Institute, Journal 187: 319–37.
1919b  The role of temperature in the determination of the transpiring power of leaves by hygrometric paper. Plant World 22: 172–80.
1919c  A thermo-electrical method for the determination of leaf temperature. Plant World 22: 100–4.
1923  Seasonal changes in the water relations of desert plants. Ecology 4: 266–92.
1924  Factors governing seasonal changes in transpiration of *Encelia farinosa*. Botanical Gazette 77: 432–39.
SHREVE, F.
1910a  The coastal deserts of Jamaica. Plant World 13: 129–34.
1910b  The rate of establishment of the giant cactus. Plant World 13: 235–40.
1911a  Establishment behavior of the Palo Verde. Plant World 14: 289–96.
1911b  The influence of low temperatures on the distribution of the giant cactus. Plant World 14: 136–46.
1912  Cold air drainage. Plant World 15: 110–15.
1913  A guide to the salient physical and vegetational features of the vicinity of Tucson, Arizona. International Phytogeographic excursion in America. (Privately printed). 11 pp.
1914a  Acidity in cacti. [a review of Spoehr (1913), *q.v.*] Plant World 17: 194–95.
1914b  The direct effects of rainfall on hygrophilous vegetation. Journal of Ecology 2(2): 82–98. (1 plate)
1914c  A montane rain-forest: a contribution to the physiological plant geography of Jamaica. Carnegie Institution of Washington, Publication 199. 110 pp.
1914d  The role of winter temperatures in determining the distribution of plants. American Journal of Botany 1: 194–202.
1914e  Rainfall as a determinant of soil moisture. Plant World 17: 9–26.
1915  The vegetation of a desert mountain range as conditioned by climatic factors. Carnegie Institution of Washington, Publication 217. 112 pp.
1916a  Excursion impressions. San Diego Society of Natural History, Transactions 2(3): 79–83.
1916b  The weight of physical factors in the study of plant distribution. Plant World 19: 53–67.
1917a  The density of stand and rate of growth of Arizona yellow pine as influenced by climatic conditions. Journal of Forestry 15: 695–707.

1917*b* The establishment of desert perennials. Journal of Ecology 5: 210–16.

1917*c* A map of the vegetation of the United States. Geographical Review 3: 119–25.

1917*d* The physical control of vegetation in rain-forest and desert mountains. Plant World 20: 135–41.

1918 The Jamaican filmy ferns. American Fern Journal 8: 65–71.

1919 A comparison of the vegetational features of two desert mountain ranges. Plant World 22: 291–307.

1922 Conditions indirectly affecting vertical distribution on desert mountains. Ecology 3: 269–74.

1924*a* Across the Sonoran Desert. Torrey Botanical Club, Bulletin 51: 283–93.

1924*b* The growth record in trees. p. 89–116. *In* Growth in trees and massive organs of plants. Carnegie Institution of Washington, Publication 350.

1924*c* Soil temperature as influenced by altitude and slope exposure. Ecology 5: 128–36.

1925 Ecological aspects of the deserts of California. Ecology 6:93–103.

1926*a* The Desert Laboratory. Progressive Arizona 3(4): 10–11, 40.

1926*b* The desert of northern Baja California. Torrey Botanical Club, Bulletin 53: 129–36.

1927*a* The physical conditions of a coastal mountain range. Ecology 8: 398–414.

1927*b* Soil temperatures in redwood and hemlock forests. Torrey Botanical Club, Bulletin 54: 649–56.

1927*c* The vegetation of a coastal mountain range. Ecology 8: 27–44.

1929 Changes in desert vegetation. Ecology 10: 364–73.

1931*a* The cactus and its home. Williams and Wilkins Co., Baltimore. 195 pp.

1931*b* The Desert Laboratory of the Carnegie Institution of Washington. Collecting Net 6: 145–47.

1931*c* Die Fouquieraceen, *Larrea tridentata* Cav., *Carnegiea gigantea* Britton et Rose. Die Pflanzenareale 3(1): 3–8. Maps 4–6.

1931*d* Physical conditions in sun and shade. Ecology 12: 96–104.

1934*a* The desert and its life. Carnegie Institution of Washington, News Service Bulletin 3(16): 113–20.

1934*b* Rainfall, runoff and soil moisture under desert conditions. Association of American Geographers, Annals 24: 131–56.

1934*c* Vegetation of the northwestern coast of Mexico. Torrey Botanical Club, Bulletin 61: 373–80.

1934*d* The problems of the desert. Scientific Monthly 199–209.

1935*a* A desert by the sea. Carnegie Institution of Washington, News Service Bulletin 3(26): 205–10.

1935*b* The human ecology of Baja California. Association of Pacific Coast Geographers, Year Book 1: 9–13.

1935*c* The longevity of cacti. Cactus and Succulent Journal 7: 66–68.

1935*d* Nordamerikanische Wüstenpflanzen, II. Die Pflanzenareale 4(3): 17–24. Maps 21–26.

SHREVE, F., *continued*
1936*a* Plant life of the Sonoran desert. Carnegie Institution of Washington, Supplementary Publication 22. 19 pp. Also *in* Scientific Monthly (1936), 42: 195–213.
1936*b* The transition from desert to chaparral in Baja California. Madroño 3: 257–64.
1937*a* Discussion of influence of vegetation on land-water relationships. *In* Headwaters Control and Use, pp. 95–100. U.S. Soil Conservation Service, Washington, D. C.
1937*b* Lowland vegetation of Sinaloa. Torrey Botanical Club, Bulletin 64: 605–13.
1937*c* Plants of the sand. Carnegie Institution of Washington, News Service Bulletin 4(10): 91–96.
1937*d* The vegetation of the Cape region of Baja California. Madroño 4: 105–13.
1938    The sandy areas of the North American Desert. Association of Pacific Coast Geographers, Yearbook 4: 11–14.
1939    Observations on the vegetation of Chihuahua. Madroño 5: 1–13.
1940    The edge of the desert. Association of Pacific Coast Geographers, Yearbook 6: 6–11.
1942*a* The desert vegetation of North America. Botanical Review 8: 195–246.
1942*b* Grassland and related vegetation in northern Mexico. Madroño 6: 190–98.
1942*c* The life forms and flora of the North American desert. American Scientific Congress, 8th, Washington, D.C., 1940, Proceedings 3: 125–32.
1942*d* Vegetation of Arizona. *In* T. H. Kearney and R. H. Peebles, Flowering plants and ferns of Arizona. U.S. Department of Agriculture, Miscellaneous Publication 423: 10–23.
1942*e* The vegetation of Jamaica. Chronica Botanica 7: 164–66.
1944    Rainfall of northern Mexico. Ecology 25: 105–11.
1945    The saguaro *(Cereus giganteus)*, Cactus Camel of Arizona. National Geographic Magazine 88: 695–704.
1951    Vegetation of the Sonoran Desert. Carnegie Institution of Washington, Publication 591. 192 pp.
——— and A. L. HINCKLEY
1937    Thirty years of change in desert vegetation. Ecology 18: 463–78.
——— and T. D. MALLERY
1933    The relation of caliche to desert plants. Soil Science 35: 99–112.
——— and W. V. TURNAGE
1936    The establishment of moisture equilibrium in soil. Soil Science 41: 351–55.
——— and I. L. WIGGINS
1964    Vegetation and Flora of the Sonoran Desert. Stanford University Press, Stanford. 1740 pp.
SIMPSON, B. B.
1977    Mesquite: its biology in two desert ecosystems, US/IBP Synthesis Series 4, Dowden, Hutchinson and Ross, Stroudsburg, Penn. 250 pp.

SINCLAIR, J. C.
1922    Temperature of the soil and air in a desert. Monthly Weather Review 50: 142–44.
SPALDING, E. S.
1905    Mechanical adjustment of the suaharo *(Cereus giganteus)* to varying quantities of stored water. Torrey Botanical Club, Bulletin 32: 57–68.
SPALDING, V. M.
1904    Biological relations of certain desert shrubs. I: The creosote bush *(Covillea tridentata)* in its relation to water supply. Botanical Gazette 38: 122–38.
1905a  Economy in irrigation. Popular Science Monthly 67: 684–86.
1905b  Soil water in relation to transpiration. Torreya 5: 25–27.
1906a  Absorption of atmospheric moisture by desert shrubs. Torrey Botanical Club, Bulletin 33: 367–75.
1906b  Biological relations of desert shrubs. II: Absorption of water by leaves. Botanical Gazette 41: 262–82.
1907a  Notes on the vegetation of Box Cañón. Plant World 10: 11–17.
1907b  Spring flowers of the Arizona desert. Plant World 10: 63–64.
1907c  Suggestions to plant collectors. Plant World 10: 40.
1909a  Cultivated plants in the arid southwest. Plant World 12: 18–21.
1909b  [A discussion of D. Griffiths, The "spineless" prickly pears, U.S.-D.A., Bureau of Plant Industry, Bulletin 140, 1909, 24 pp.] Plant World 12: 89–94.
1909c  Distribution and movements of desert plants. Carnegie Institution of Washington, Publication 113. 144 pp.
1909d  The western edge of the Colorado Desert. Plant World 11: 208–15.
1910    Plant associations of the Desert Laboratory domain and adjacent valley. Plant World 13: 31–42, 56–66, 86–93.
SPOEHR, H. A.
1911    The relation between photosynthesis of carbon dioxide and nitrate reduction. Science 34: 63–64.
1913    Photochemische vorgaenge bei der diurnalen entsaeuerung der succulenten. Biochemische Zietschrift 57: 95–111.
1915    Variations in respiratory activity in relation to sunlight. Botanical Gazette 59: 366–86.
1916    The theories of photosynthesis in the light of some new facts. Plant World 19: 1–16.
1917    The pentose sugars in plant metabolism. Plant World 20: 365–79.
1919a  The carbohydrate economy of cacti. Carnegie Institution of Washington, Publication 287. 79 pp.
1919b  The development of conceptions of photosynthesis since Ingen-Housz. Scientific Monthly 9: 32–46.
1922    Photosynthesis and the possible use of solar energy. Journal of Industrial and Engineering Chemistry 14(12): 1,142–45.
1923    The reduction of carbon dioxide by ultraviolet light. American Chemical Society, Journal 45: 1,184–87.
1924    The oxidation of carbohydrates with air. American Chemical Society, Journal 46: 1,494–1,502.

SPOEHR, H. A., *continued*
1927    Photosynthesis. American Chemical Society, Monograph Series. Chemical Catalog Co., Inc., New York. 393 pp.
1967    Form, forces and function in plants. *In* C. P. Haskins, ed., The Search for Understanding. p. 191–209.
——— and J. M. McGEE
1923*a* Investigations in photosynthesis; an electrometric method of determining carbon dioxide. Industrial and Engineering Chemistry 16(2): 128–30.
1923*b* Studies in plant respiration and photosynthesis. Carnegie Institution of Washington, Publication 325. 98 pp.
1924    Absorption of carbon dioxide the first step in photosynthesis. Science, n.s. 59: 513–14.
1967    Form, function and forces in plants. *In* Haskins, C. P., ed., *The Search for Understanding.* pp. 191–209.
SPOEHR, H. A., and H. S. MILLER
1956    Essays on science by Herman Augustus Spoehr, a selection of his works. Stanford University Press. 220 pp.
——— and J. H. C. SMITH
1926    Studies on atmospheric oxidation. I: Oxidation of glucose and related substances in the presence of sodium ferro-pyrophosphate. II: Kinetics of the oxidation with sodium ferro-pyrophosphate. American Chemical Society, Journal 48: 236–48. 107–12.
——— and P. C. WILBUR
1926    The effect of disodium phosphate on $d$-glucose and $d$-fructose. Journal of Biological Chemistry 69(2): 421–34.
STANDLEY, P. C.
1920–26 Trees and shrubs of Mexico. U.S. National Herbarium Contributions 23: 1–1,721.
SYKES, G.
1912    A journey in the Libyan desert. American Geographical Society, Bulletin 44(10): 721–45.
1914*a* Agriculture in the Nile Valley. Plant World 17: 69–75.
1914*b* Geographical features of the Cahuilla Basin. *In* D. T. MacDougal and Collaborators, The Salton Sea: A study of the geography, the geology, the floristics, and the ecology of a desert basin. Carnegie Institution of Washington, Publication 193: 13–20.
1915*a* How California got its name. Out West, Los Angeles, California 41(6): 225–30.
1915*b* The isles of California. American Geographical Society, New York, Bulletin 47(10): 745–61.
1915*c* The mythical straits of Anian. American Geographical Society, Bulletin 47(3): 161–72. 10 maps.
1915*d* The reclamation of a desert. Geographical Journal, London 46(6): 447–57.
1931    Rainfall investigations in Arizona and Sonora by means of long-period rain gauges. Geographical Review 21: 229–33.

1937a  The Colorado Delta. Carnegie Institution of Washington, Publication 460. 193 pp.
1937b  Delta, estuary, and lower portion of the channel of the Colorado River 1933 to 1935. Carnegie Institution of Washington, Publication 480. 70 pp.
1938   End of a great delta. Pan-American Geologist 69: 241–48.
1939a  Regolith of the desert. Pan-American Geologist 71: 347–58.
1939b  Rio Santa Cruz of Arizona; paradigm desert stream-way. Pan-American Geologist 72: 81–92.
1944   A Westerly Trend. Arizona Pioneer Historical Society. 325 pp.
THORNBER, J. J.
1909a  Vegetation groups of the Desert Laboratory domain. Plant World 12: 289–93.
1909b  Vegetation groups of the Desert Laboratory domain. In Spalding, 1909, Distribution and movements of desert plants. Carnegie Institution of Washington, Publication 113.
TOUMEY, J. W.
1905   Notes on the fruits of some species of Opuntia. Torrey Botanical Club, Bulletin 32: 235–39.
TOWER, W. L.
1907   An investigation of evolution in chrysomelid beetles of the genus Leptinotarsa. Carnegie Institution of Washington, Publication 48. 320 pp.
1910   The determination of dominance and the modification of behavior in alternative (Mendelian) inheritance, by conditions surrounding or incident upon the germ cells at fertilization. Biological Bulletin 18(6): 285–352.
1919   The mechanism of evolution in Leptinotarsa. Carnegie Institution of Washington, Publication 236. 384 pp.
TURNAGE, W. V.
1937a  Nocturnal surface-soil temperatures, air temperatures, and ground inversions in southern Arizona. Monthly Weather Review 65: 189–90.
1937b  Note on accuracy of soil thermographs. Soil Science 43: 475–76.
1939   Desert subsoil temperatures. Soil Science 47: 195–99.
——— and A. L. HINCKLEY
1938   Freezing weather in relation to plant distribution in the Sonoran Desert. Ecological Monographs 8: 529–50.
——— and T. D. MALLERY
1941   An analysis of rainfall in the Sonoran Desert and adjacent territory. Carnegie Institution of Washington, Publication 529. 45 pp.
——— and E. B. SHREVE
1939   Note on atmospheric aridity. Ecology 20: 107–9.
TURNER, R. M.
1963   Growth in four species of Sonoran Desert trees. Ecology 44: 760–65.
———, et al.
1966   The influence of shade, soil, and water on saguaro seedling establishment. Botanical Gazette 127(2–3): 95–102.

VINSON, A. E.
1914   Variations in composition and concentration of water of Salton Sea, 1912, and 1913. *In* D. T. MacDougal and Collaborators, The Salton Sea: A study of the geography, the geology, the floristics, and the ecology of a desert basin. Carnegie Institution of Washington, Publication 193: 47–48.

WALTER, H.
1931   Die Kyroscopische Bestimmung des osmotischen Wertes bie Pflanzen. Handbuch der Biologischen Arbetismethoden 11(4:2): 353–71.
1932   Die wasservarhältnisse an verschiedenen standorten in humiden und ariden gebieten. Botanisches Centralblatt, Biehefte, Drudefestschr. 1932. 49, Erg.-Bd., 495–514.

WEAVER, JOHN
1919   Ecological relations of roots. Carnegie Institution of Washington Publication 286.

WHITTAKER, R. H., and W. A. NIERING
1964   Vegetation of the Santa Catalina Mountains, Arizona, I: Ecological classification and distribution of species. Arizona Academy of Science, Journal 3(1): 9–34.

WIGGINS, I. L.
1934   A report on several species of *Lycium* from the southwestern deserts. Stanford University, Dudley Herbarium, Contribution 1(6): 195–209.
1935   An extension of the known range of the Mexican bald cypress. Torreya 35: 65–67.
1937a   Effects of the January freeze upon the pitahaya in Arizona. Cactus and Succulent Society, Journal 8: 171.
1937b   Notes on the habitat and distribution of *Grusonia wrightiana*. Cactus and Succulent Society, Journal 8: 134–35.
1939a   Distributional notes on and a key to the species of *Cheilanthes* in the Sonoran Desert and certain adjacent regions. American Fern Journal 29: 59–69.
1939b   Folklore and fact of medicinal plants in rural Mexico. New York Botanical Garden, Journal 40: 176–79.
1940a   Bizarre trees and other strange plants from Lower California. New York Botanical Garden, Journal 41: 201–8.
1940b   New and poorly known species of plants from the Sonoran Desert. Stanford University, Dudley Herbarium, Contribution 3(3): 65–84.
1940c   Taxonomic notes on the genus *Dalea* Juss. and related genera as represented in the Sonoran Desert. Stanford University, Dudley Herbarium, Contribution 3(2): 41–55.
1940d   Yellow pines and other conifers observed in Lower California. New York Botanical Garden, Journal 41: 267–69.
1942   *Acacia angustissima* (Mill.) Kuntze and its near relatives. Stanford University, Dudley Herbarium, Contribution 3(7): 227–39.
1943   Two new plants from the San Felipe Desert, Baja, California, Mexico. Stanford University, Dudley Herbarium, Contribution 3(8): 285–88.

1944a Collecting ferns in northwestern Mexico. American Fern Journal. 34: 37–49.

1944b The genus *Drymaria* in, and adjacent to, the Sonoran Desert. California Academy of Science, Proceedings. ser. 4, 25: 189–213.

1944c Notes on the plants on northern Baja California, Stanford University, Dudley Herbarium, Contribution 3(9): 289–305.

1944d Notes on the plants of northern Baja California. Contributions from the Dudley Herbarium. 3(9):289–312.

1960  Investigations in the natural history of Baja California. Proceedings of the California Academy of Sciences, 4th series 30(1): 1–45.

1969  Observations on the Vizcaíno Desert and its biota. California Academy of Sciences, Proceedings, 4th series 36(11): 317–49.

1980  Flora of Baja, California. Stanford University Press. 1025 pp.

——— and R. C. ROLLINS

1943  New and noteworthy plants from Sonora, Mexico. Stanford University, Dudley Herbarium, Contribution 3(8): 266–84.

WILDER, J. C.

1967  The years of a Desert Laboratory. Journal of Arizona History 8: 179–99.

WILDER, J. D.

1930  A modified form of non-absorbing valve for porous cup atmometers. Science 71: 101–3.

# Acknowledgments

A slowly evolving manuscript project accumulates many debts for the author. In 1975, the manuscript of my original history of the Desert Botanical Laboratory was completed, having been made possible by a grant from the Carnegie Institution of Washington, D.C. This draft is in the archives of the Carnegie Institution and in Special Collections at the University of Arizona.

The present book is the outgrowth of that manuscript and of several more years of research and writing generously supported by facilities and services of the University of Arizona.

Material was contributed and review assistance given for the original project by Margaret Shreve Conn, daughter of Forrest Shreve; Howard Scott Gentry, Desert Botanical Laboratory, Phoenix, Arizona; Robert R. Humphrey, Emeritus Professor of Range Management, University of Arizona; T. Dwight Mallery, retired former Desert Laboratory staff member; Hortense Spoehr Miller, daughter of Herman Augustus Spoehr; Glenton Sykes, retired, son of Godfrey Sykes; Raymond M. Turner, Research Botanist, U.S. Geological Survey; Ira L. Wiggins, Professor Emeritus, Department of Biological Sciences, Stanford University, co-author with Forrest Shreve of *Vegetation and Flora of the Sonoran Desert.*

Others who assisted in the preparation of that first draft were Patricia Paylore, Mary Jane Michael, Elena Turano, Cecilia Chavarin, and Mary Chavez, all associated with the Office of Arid Lands Studies, University of Arizona; and Rose Samardzich, secretary, College of Earth Sciences, University of Arizona.

The present book owes much to Jack Johnson, Director of the Office of Arid Lands Studies, who made facilities and personnel of the Office of Arid Lands Studies available to me for several months' work.

The revision of the original manuscript was done with substantial assistance from Rebecca Staples, who made major suggestions for content and arrangement of material, and carried out the many editorial tasks essential to the preparation of the manuscript for submission to a publisher. Without her help and guidance this book, in the present form, would not have been written; she therefore merits the greatest possible appreciation for her contributions.

I owe much to my wife, Rose, who helped in a million little ways. She shared with me the many hopes and disappointments incidental to preparing a book of this kind, and checked measurements and metric conversion which have a habit of becoming erroneous during manuscript preparation.

Finally, I owe a great deal to the University of Arizona Press and its staff for their acceptance of this manuscript and the subsequent production activities essential to its publication.

W. G. McG.

# Index